Simulation Foundations, Methods and Applications

For further volumes:
http://www.springer.com/series/10128

Ercan M. Dede · Jaewook Lee
Tsuyoshi Nomura

Multiphysics Simulation

Electromechanical System Applications and Optimization

Springer

Authors
Ercan M. Dede
Toyota Research Institute
of North America
Ann Arbor, MI
USA

Tsuyoshi Nomura
Toyota Central R&D Labs.
Nagakute, Aichi
Japan

Jaewook Lee
Korea Aerospace University
Goyang-si, Kyonggi-do
Korea, Republic of (South Korea)

Series editor
Prof. Emeritus Louis G. Birta
University of Ottawa
Ottawa, ON
Canada

ISSN 2195-2817 ISSN 2195-2825 (electronic)
ISBN 978-1-4471-5639-0 ISBN 978-1-4471-5640-6 (eBook)
DOI 10.1007/978-1-4471-5640-6
Springer London Heidelberg New York Dordrecht

Library of Congress Control Number: 2014938471

Printed on acid-free paper

Springer is part of Springer Science+Business Media (www.springer.com)

Foreword

It was 1986, at the Technical University of Denmark, when Prof. Martin Bendsøe and I were working on 'topology' optimization in structural design by finding an optimal distribution of a linearly elastic porous microstructure defined by infinitely many very small-scale rectangular holes, which are appropriately rotated. To deal with a linearly elastic porous material defined by infinitely many small-scale rectangular holes (characterized by width, a, and height, b, in a unit cell with rotation, θ), we applied the homogenization method to calculate the equivalent macro-scale linearly elastic constitutive relation, $\sigma = \mathbf{C}^H(a, b, \theta)\varepsilon$. Instead of rectangular holes, square holes were assumed characterizing size, a, in the unit cell, and the elasticity matrix, $\mathbf{C}^H(a, b, \theta)$, was simplified to $\mathbf{C}^H(a, \theta)$. If rotation of micro-scale holes is not considered, then this problem is simplified even further as $\mathbf{C}^H(a)$. If $a = 1$, then the unit cell is completely occupied with a hole, that is, it is equivalent to no material in the macro-scale 'porous' structure. On the other hand, if $a = 0$, then the unit cell is completely occupied with elastic material, that is, it represents solid material in the macro-scale 'porous' material. Shape and topology of a solid structure, then may be defined by the portion where $a < 1$ in the unit cell. In other words, topology and shape optimization problems are transferred to finding the optimal distribution of porosity, that is, the optimal distribution of a solid material. This idea was published in the paper Bendsøe and Kikuchi [1] and then its concept was extensively extended by Bendsøe [2], Bendsøe and Sigmund [3].

In these books, we can find applications not only in elastic structures, but in heat conduction and also in fluid mechanics. However, at present, a dependable monograph cannot be found on topology optimization for electromagnetism and its industrial applications, which are now very critical to developing various advanced electrical/magnetic devices even for automobiles.

Thus, I have encouraged the writing of such a monograph to researchers at the Toyota Research Institute of North America, Ann Arbor, Michigan, USA, where the topology optimization approach is widely taken as a design tool for many devices and structures for automobiles. It is my great pleasure to endorse this monograph to many talented graduate students and researchers in the field of electromagnetism and its related applications in design.

April 2014 Noboru Kikuchi

References

1. Bendsøe MP, Kikuchi N (1988) Generating optimal topologies in structural design using a homogenization method. Comput Methods Appl Mech Eng 71(2):197–224
2. Bendsøe MP (1995) Optimization of structural topology, shape, and material. Springer, Berlin
3. Bendsøe MP, Sigmund O (2003) Topology optimization: theory, methods, and applications, 2nd edn. Springer, Berlin

Preface

This book developed out of a collaboration by the authors at the Toyota Research Institute of North America, where multiphysics simulation and optimization is used on a daily basis for a variety of engineering studies related to electromechanical systems. Multiphysics simulation is a rapidly growing field, and the term itself is broad and may be applied to an extremely wide variety of coupled-physics problems. By nature, multiphysics simulation requires an array of technical skills in different intersecting disciplines. As such, this book aims to narrow down the topic by specifically focusing on multiphysics simulation for electromechanical systems, the original target application investigated by the authors. It is our hope that the collaborative aspects of such studies become apparent as the various technical topics throughout the book are presented.

Overview

Understanding and predicting the performance of electromechanical systems is of prime importance in the design of many of today's key products including computers, vehicles and consumer electronics. In these systems, increased efficiency and higher power density in a smaller package size is crucial. Success in design requires both analytical and numerical skills plus a foundation in mechanical and electrical engineering. Efficient analysis also necessitates an understanding of how best to build a numerical model that is accurate yet balances complexity and computational cost.

Beyond basic performance prediction, today's engineers and researchers are constantly seeking methods for optimizing complex electromechanical systems. The multiphysics aspects of these systems present constant challenges in terms of how best to arrive at an 'optimal solution.' Many optimization techniques exist, although the use of structural topology optimization is emphasized herein along with some aspects of discrete parameter optimization.

Organization and Features

Accordingly, this book highlights a unique combination of numerical tools including numerous strategies for handling the aforementioned simulation challenges. In Chap. 1, the concept of design via simulation is introduced along with the role of multiphysics simulation in today's engineering environment. The importance of structural optimization techniques in the design and development of electromechanical systems is additionally discussed. From there, an overview of the physics commonly involved with electromechanical systems is provided (Chap. 2) for applications such as electronics, magnetic components, radio frequency components, actuators and motors. Governing equations for the simulation of related multiphysics problems are reviewed in Chap. 3, while the relevant topology optimization and parametric size analysis methods for electromechanical systems are outlined in Chap. 4. Several multiphysics simulation and optimization example studies in both two and three-dimensions are then described in detail throughout Chap. 5. Extensions to new topics are suggested in Chap. 6. Sample numerical code for a related electro-thermal topology optimization example is provided in the appendix in Chap. 7.

A challenge in writing a book of this nature on the topic of multiphysics simulation is the preparation of the engineering nomenclature used for various physical constants, state variable, functions, etc. Specifically, the governing equations for the multitude of separate physical processes are often described using the same symbols for different variables. As such, every effort has been made to provide a comprehensive list of nomenclature with distinct variable usage wherever possible.

In addition to the above features, extensive references are provided at the end of each chapter. These references are related to prior research on multiphysics simulation and optimization methods, techniques, and application studies.

Target Audiences

It is our hope that the content presented in this book will serve as a reference for industry and academic researchers and engineers in the field of advanced electromechanical system design. The topics in this book are appropriate for undergraduate and graduate level students, although many of the design examples may be of interest to anyone curious about the unique design solutions that arise when optimization methods are coupled with multiphysics simulation strategies.

Acknowledgments

We would like to thank our colleagues at the Toyota Research Institute of North America (TRINA), Korea Aerospace University, and Toyota Central Research and Development Labs (TCRDL), for their support in the completion of this project.

Professor Emeritus Louis G. Birta, Series Editor for Springer's Simulation Foundations, Methods and Applications book series, deserves special thanks for the invitation to write this manuscript. Professor Gregory M. Hulbert from the University of Michigan plus Dr. Danil Prokhorov and Koji Shiozaki from TRINA are also acknowledged for their kind encouragement in undertaking this project. Our appreciation goes to Satoru Sasaki from Toyota Motor Corporation for his review of the text and technical feedback, Dr. Yan Liu from Toyota Technical Center for discussions on thermal-fluid system design, and Prof. Xiaoping Qian from the University of Wisconsin-Madison for interesting conversations on optimization and CAE.

We further thank Dr. Kazuo Sato and Dr. Kunitoshi Nishikawa of TCRDL for providing the opportunity and support for this collaborative work, Paul Schmalenberg from TRINA, Dr. Hisayoshi Fujikawa and Makoto Ohkado from TCRDL for prototyping and measurement of the topology optimized microstrip coupler, Prof. Shintaro Yamasaki of Osaka University for technical feedback on the level set method and contribution of figures, Prof. Tatsuya Kashiwa of Kitami Institute of Technology and Prof. Mohamed Bakr of McMaster University for discussions on adjoint analysis with time domain solvers, Prof. Sang Won Yoon of Hanyang University for the topology optimized DBC prototyping, Keiichi Shimaoka and Takashi Ozaki of TCRDL for inviting an author to participate in their MEMS project, Dr. Tadayoshi Matsumori and Dr. Atsushi Kawamoto of TCRDL for discussions on filtering methods and vectorial topology optimization, and Prof. Shinji Nishiwaki of Kyoto University for leading guidance in the field of topology optimization, especially in Asia.

We also would like to thank Prof. Yoon Young Kim of Seoul National University, Prof. Seonho Cho of Seoul National University, Prof. Jeonghoon Yoo of Yonsei University, Prof. Seungjae Min of Hanyang University, and Prof. Gil Ho Yoon of Hanyang University for their support and advice on topology optimization. Additionally, we wish to express special thanks to Prof. Joo Ho Choi and all of the Professors at Korea Aerospace University for their encouragement in completing this project.

A very special acknowledgment is given to Professor Noboru Kikuchi for his kind support of this project and pioneering research in the field.

Finally, this book would not be possible without the encouragement and strong support of our families and friends.

Ann Arbor, April 2014 Ercan M. Dede
Goyang Jaewook Lee
Tsuyoshi Nomura

Contents

Acronyms

2-D	Two-dimensional
3-D	Three-dimensional
ABC	Absorbing boundary condition
AC	Alternating current
ALE	Arbitrary Lagrangian–Eulerian
AVM	Adjoint variable method
BC	Boundary condition
CAD	Computer-aided design
CAE	Computer-aided engineering
CFD	Computational fluid dynamics
CTE	Coefficient of thermal expansion
CVW	Coulomb Virtual Work
DARPA	Defense Advanced Research Projects Agency
DBC	Direct bonded copper
DC	Direct current
DOF	Degrees-of-freedom
DRA	Dielectric resonator antenna
EMC	Electromagnetic compatibility
EMI	Electromagnetic interference
FDTD	Finite difference-time domain
FEA	Finite element analysis
FE-BI	Finite element-boundary integration
FEM	Finite element method
GA	Genetic algorithm
HPC	High performance computing
HV	Hybrid vehicle
IC	Integrated circuit
IPMSM	Interior permanent magnet synchronous motor
LSM	Level set method
MEMS	Microelectromechanical systems
MMA	Method of Moving Asymptotes
MST	Maxwell Stress Tensor
NASA	National Aeronautics and Space Administration
OC	Optimality criteria

PCB	Printed circuit board
PDE	Partial differential equation
PEC	Perfect electric conductor
PM	Permanent magnet
PML	Perfectly matched layer
PSO	Particle swarm optimization
RF	Radio frequency
RMS	Root mean square
RPM	Revolutions per minute
SAR	Specific absorption rate
SAW	Surface acoustic wave
SIMP	Solid isotropic material with penalization
SLP	Sequential linear programming
SRM	Switched reluctance motor
TBC	Transition boundary condition
TEM	Transverse electric and magnetic
TIM	Thermal interface material
US	United States

Symbols

Scalar Quantities

A_s	Surface area
B	Magnitude of magnetic flux density
B_{r_PM}	Permanent magnet strength
B_{rx}	x-direction component of residual magnetic flux density
B_{ry}	y-direction component of residual magnetic flux density
C	Curie constant
C_p	Specific heat capacity
D	Extended design domain
E	Elastic (or Young's) modulus
F	Force
f	Frequency
f_i	Inclusion volume fraction
g	Behavior constraint
H	Magnitude of magnetic field
h	Surface convection coefficient
h_c	Channel height
h_t	Heaviside function transition bandwidth
I, i	Electric current
J	Electric current density
k	Thermal conductivity
k_h	Hysteresis coefficient
L	Inductance
l	Mechanical compliance
M_w	Molecular weight
N	Nodal shape function
N_c	Number of coils
P	Pressure
p	Penalization parameter
P_r	Number of rotor poles
P_s	Number of stator poles
$\mathscr{P}$	Power

Q	Volumetric power density
q	Convex interpolation tuning parameter
q''	Heat flux
R	Electrical resistance
R_f	Helmholtz filter radius
R_{th}	Thermal resistance
R''_{th}	Unit thermal resistance
$R_{\text{th(cnv)}}$	Convective thermal resistance
Re	Reynolds number
$\mathscr{R}$	Residual
S	Scattering parameter
s	Adaptive scaling factor
T	Temperature
t	Time
$\mathscr{T}$	Torque
u	Displacement
V	Voltage drop
v	Volume
v_e	Element volume
W	Energy or work
W_{co}	Co-energy
W_{mag}	Magnetic energy
w	Objective function weighting value
w_c	Channel width
Z_s	Surface impedance
α	Coefficient of thermal expansion
α_i	Inclusion aspect ratio
$\tilde{\alpha}$	Inverse permeability
χ	Fluid magnetic susceptibility
$\delta\varepsilon$	Virtual strain
δU	Virtual strain energy
δW	Virtual work
ΔP	Pressure drop
ϵ	Electric permittivity
ϵ_o	Electric permittivity of free space
ε_b	Bandwidth for smoothed surface convection coefficient function
η	Fluid dynamic viscosity
γ	Optimization design (density) variable
λ_{link}	Magnetic flux linkage
μ	Magnetic permeability
μ_o	Magnetic permeability of free space
μ_r	Relative magnetic permeability
μ_{r_mf}	Magnetic fluid permeability

ν	Poisson's ratio
Ω	Domain
Ω_d	Design domain
Ω_m	Material domain
ω	Angular frequency or velocity
Φ	Electric scalar potential
Φ_B	Magnetic flux
ρ	Density
ρ_e	Electric resistivity
ρ_q	Electric charge density
ς	Electric conductivity
θ	Inclusion angle
θ_r	Rotor angle
ζ	Channel aspect ratio
Vector, Matrix, or Tensor Quantities	
$\mathbf{A}$	Magnetic vector potential
$\mathbf{B}$	Magnetic flux density
$\mathbf{B}_r$	Residual magnetic flux density
$\mathbf{B}_e$	Array of derivatives of element nodal shape functions
$\mathbf{C}$	Stiffness tensor
$\mathbf{D}$	Electric displacement
$\mathbf{E}$	Electric field
$\mathbb{E}$	Electric field variable
$\hat{\mathbf{e}}$	Target unit vector
$\mathbf{F}$,$\mathbf{f}$	Force (global, element)
$\mathbf{H}$	Magnetic field
$\mathbf{J}$	Electric current density
$\mathbf{J}_e$	External current density
$\mathbf{J}_{\text{eddy}}$	Eddy current density
$\mathbf{K}$,$\mathbf{k}$	Stiffness matrix (global, element)
$\hat{\mathbf{n}}$	Normal unit vector
$\mathbf{Q}$	Magnetic load vector
$\mathbf{q}$	Heat flux vector
$\mathbf{t}$	Surface load vector
$\hat{\mathbf{t}}$	Tangential unit vector
$\mathbf{U}$,$\mathbf{u}$	Displacement (global, element)
$\mathbf{v}$	Velocity
$\mathbf{x}$	Position
$\delta\mathbf{u}$	Virtual displacement
ε	Strain
$\hat{\lambda}$	Lagrange multipliers (adjoint variables)
σ	Stress

Functions

F_o	Optimization objective function
F_c	Convergence function
H	Heaviside function
$\tilde{\mathsf{H}}$	Regularized Heaviside function
L	Lagrangian
ϕ	Level set function
ψ	Scalar function
$\tilde{\psi}$	Filtered scalar function

Subscripts

avg	Average
d	Distributed
F,f	Fluid
in	Inlet
init	Initial
l	Lower
M	Magnetic
max	Maximum
min	Minimum
o	Reference value
out	Outlet
S	Structural
s	Solid
T	Thermal
u	Upper

Chapter 1
Introduction

Starting in the mid-1950s and progressing into the following two decades, the mathematical framework behind the finite element method was developed by researchers due to the potential of the method as a revolutionary tool in the analysis and design of civil and aerospace structures [17]. Building off such early work [35], and under financial support from the US federal government, NASTRAN®[1] finite element analysis (FEA) software was developed by NASA in conjunction with industry for aerospace applications. Today, FEA is a highly developed and commercially available computational approach that allows for the estimation of the response of structures that are subjected to multiple physical processes.

In this chapter, the role of FEA in design via simulation is explored and the importance of this approach for advanced electromechanical design is emphasized. Representative differences between single and multiphysics simulations are presented in the context of an example electromechanical system. Common challenges associated with multiphysics simulation are then described to motivate the discussion of the various computational tools presented throughout this book. An introductory overview of the selected optimization techniques that are presented later on in the text is additionally provided.

1.1 Design via Simulation

The use of FEA in the early stages of a product design process has become common across multiple industries including automotive, aerospace, and consumer electronics. Designing complex systems via simulation can lead to more efficient designs and significant reduction in product development cycle times as evidenced by recent aircraft structural design studies. In [14], the authors describe the use of advanced numerical topology, shape, and size optimization techniques to generate efficient stress and stability designs for aircraft leading edge droop nose ribs. This example

[1] NASTRAN® is a registered trademark of the National Aeronautics Space Administration.

E. M. Dede et al., *Multiphysics Simulation,* Simulation Foundations, Methods and Applications, DOI: 10.1007/978-1-4471-5640-6_1,

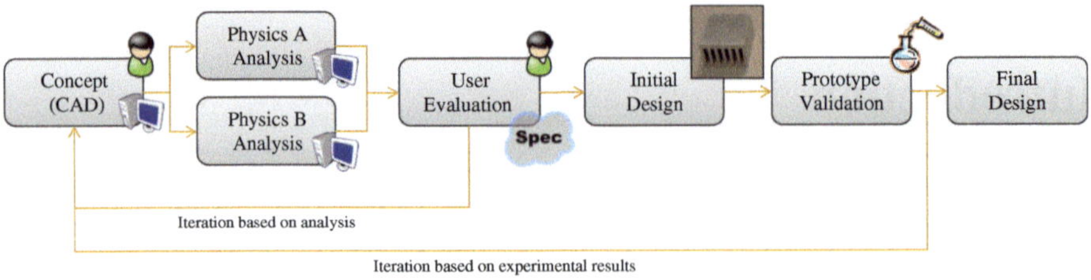

Fig. 1.1 A design cycle process flow utilizing FEA for performance evaluation

highlights the prospective role of High Performance Computing (HPC) in analyzing large computational models, where engineers can design better products in shorter time-frames with significantly less prototype testing and manual optimization effort.

The development of advanced computer aided engineering (CAE) tools for the design of future ground vehicle systems has also recently been identified as a key enabling technology by the US Defense Advanced Research Projects Agency (DARPA) as part of their Fast Adaptable Next-Generation Ground Vehicle program. In this program, DARPA seeks to establish 'fab-less design' capabilities through a suite of integrated simulation and modeling techniques [6]. The final objective of this ambitious project is again to develop highly optimized vehicles via simulation with significantly reduced development cycle times and cost.

A third example of the evolving design via simulation trend may be found in modern marine propulsion systems, where commercial pressures have forced companies to rely less on traditional model tests and instead implement computational design methods coupled with validation tests [23]. Once more, efficient and reliable CAE tools are enabling the effective design and optimization of these systems via analysis with experimental testing of the final concept relegated to the closing stages of the development program.

While the three aforementioned examples are focused on large-scale aerospace, ground, and naval vehicle systems, respectively, the same design via simulation principles may be applied at much smaller scales for modern micro-electronics as well as future electromechanical motors and actuators. In these applications, smaller physical size translates into highly constrained packaging, where the increased energy density brings forth a unique set of design challenges that are inherently multiphysics in nature [7].

The ultimate goal in any of these applications is to reduce iterative trial-and-error design strategies and incorporate advanced modeling and design optimization techniques into the beginning stages of the design process. This change in philosophy is visualized by comparing Fig. 1.1, an example of a traditional design cycle process flow, with Fig. 1.2, a more contemporary design cycle process approach.

In the traditional design cycle process flow, Fig. 1.1, the designer or engineer synthesizes a concept after need recognition and problem definition based on established analytical methods, applied experience, or other *a priori* knowledge [25]. Today's engineer often makes use of advanced CAE tools such as FEA to analyze the response

Fig. 1.2 A design cycle process flow utilizing advanced multiphysics simulation and optimization technologies

of the synthesized system concept [24]. From there, the designer evaluates performance and iterates on the proposed concept either during the analysis or prototype evaluation stage. It is often the case that several iterations are necessary to satisfy the performance targets, where major redesigns become increasingly costly once at the prototype evaluation stage.

While the use of theoretical techniques and engineering experience is invaluable, the goal in adopting advanced simulation and optimization technologies early in the design stage of a project, Fig. 1.2, is to avoid costly iterations at later phases of the project where less design flexibility is readily available. This ideal scenario is even more imperative as electromechanical systems become more highly integrated and space constrained, since analytical models may be limited and numerous prototypes must be fabricated when adopting a purely experimental approach in order to fully understand interactions arising from multiple physical scales and processes. In view of the foregoing, the need for advanced multiphysics simulation and optimization techniques is real and continually growing.

In this book, we aim to describe a set of numerical tools that enables efficient early stage research and development concepts for product design via multiphysics simulation and optimization. These tools may then be applied to a broad variety of electromechanical system applications, and this point is emphasized through the description of several example studies that are provided in later chapters of the text.

1.2 Single Physics Versus Multiphysics Simulation

Standard applications of FEA include the computation of the deformation or temperature of a structure due to an applied mechanical or thermal load [15]. A governing partial differential equation (PDE) describes the physical system of interest and an approximate solution to this equation is found by discretizing a large structure and solving a matrix form of the problem using linear algebra techniques. Single physics FEA simulations are commonplace in today's engineering environment, and numerous commercial software programs, see for example [19], exist for analyzing a variety of physical situations.

More recently, understanding the real-world coupled response of a structure subjected to multiple physical processes has become an area of sustained interest [15, 16,

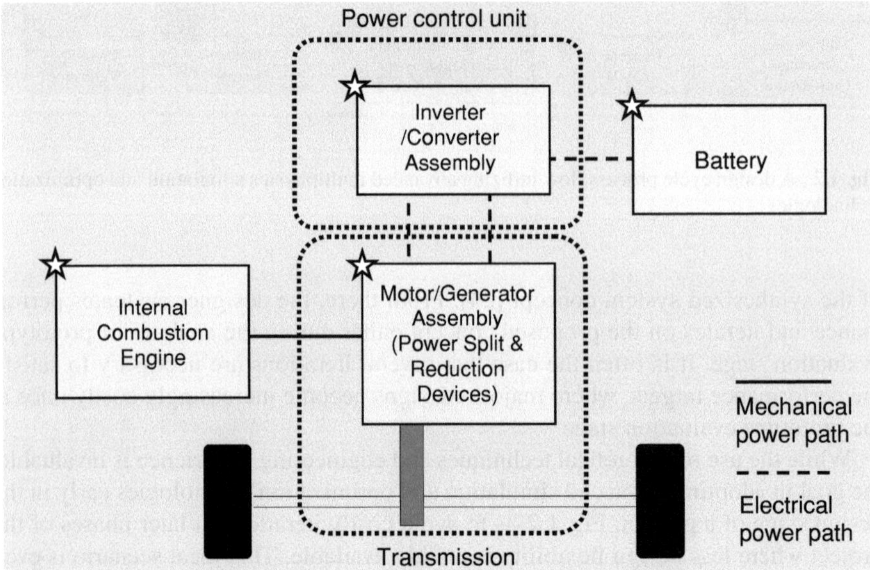

Fig. 1.3 Example hybrid vehicle drive-train schematic including traditional internal combustion engine, transmission (with motor/generator), power control unit (comprising inverter/converter), and hybrid vehicle battery. The four starred items in the figure indicate multiphysics systems involving some combination of heat transfer, fluid flow, thermal-stress, or electromagnetic phenomena. Dynamics, vibration, and control are also critical aspects to such systems and must be considered in the design process

30]. The availability of new multiphysics FEA software plus advanced workstations and HPC systems enables accurate coupled-field analysis for electronics, motors, and actuators. Here, electro-thermal, thermal-fluid, thermal-structural, and electro-magneto-thermal interactions (to name a few) are common and often lead to greater system complexity in addition to significant system efficiency and reliability challenges [15].

As an example of a multiphysics system, a representative hybrid vehicle (HV) drive-train schematic is shown in Fig. 1.3, where the starred items indicate subsystems that involve some combination of heat transfer, fluid flow, thermal stress, or electromagnetic phenomena [18, 21]. For the inverter electronics, managing steady-state and transient thermal loads plus thermally induced stresses is of prime importance, while increasing motor torque and simultaneously minimizing torque ripples is a key challenge for the electric motor [7].

This drive-train example illustrates the importance of physics coupling in advanced simulation. Specifically, it is not possible to anticipate and mitigate the stresses induced by operating an electronics system unless the designer understands the power dissipation and thermal loads that are generated by applying electrical power to the device package. Additionally, the introduction of a design variable (as in topology optimization) may also create coupling between the actual physics-based problem and the supplementary optimization problem. As a result, the numerical solution of coupled multiphysics problems is extremely important in realizing the ideal design process flow illustrated in Fig. 1.2.

1.3 Challenges of Multiphysics Simulation

As explained in [15], the primary benefits and challenges associated with multiphysics simulation stem from coupled-field analysis, where multiple physical phenomena may be modeled simultaneously. Prior to the development of multiphysics FEA software, coupled-field analysis was complicated, where the solution from one physics model and software program had to be passed to another program to solve for a second physical situation. This file transfer process required time and accuracy was not necessarily guaranteed since user errors were easily propagated with such data exchanges. Fortunately, this challenge has mostly been overcome through the creation of commercially available software packages; see for example [2, 5, 9].

Further examination of coupled-field analysis reveals that two types of coupling generally exist for multiphysics problems including direct coupling and sequential coupling [15]. In a direct coupled system, a single matrix system of equations based on all of the relevant physics is assembled and then solved. One drawback for direct coupled systems is that finding a solution can be costly in terms of required processing power and computer memory requirements. Alternatively, a segregated solver may be used for a sequentially coupled system, where the solution from a first set of field equations is passed to a second set of field equations, which is solved and then passed to a third set of equations, etc. This segregated process is repeated until a final solution is obtained. Many of the commercial multiphysics FEA packages now automatically 'suggest' an appropriate default coupling depending on the physics involved, although the user may customize these settings, as needed, based on solution time, available computing power, and numerical convergence of the problem at hand.

As discussed in the prior section, the introduction of a design variable logically creates coupling between the physics-based problem and a supplementary optimization problem. The efficiency of this coupling scheme again affects overall solution time. For topology optimization, this coupling is typically handled in a sequential fashion, where information from the FEA solution is passed to the optimizer and vice-versa until some convergence criterion is satisfied. Interestingly, there is also recent work related to transforming the design variable into an equivalent state variable in the solution of a single direct coupled physics-based problem [12]. Such an approach has inherent challenges but also has significant potential for increased useability and reduced computational cost.

Perhaps one of the greatest challenges with any multiphysics simulation is understanding all of the relevant physics involved in a particular problem and setting up a geometrically accurate model with appropriate loads and boundary conditions. Incorporating multiple physical phenomena into a simulation may require consultation

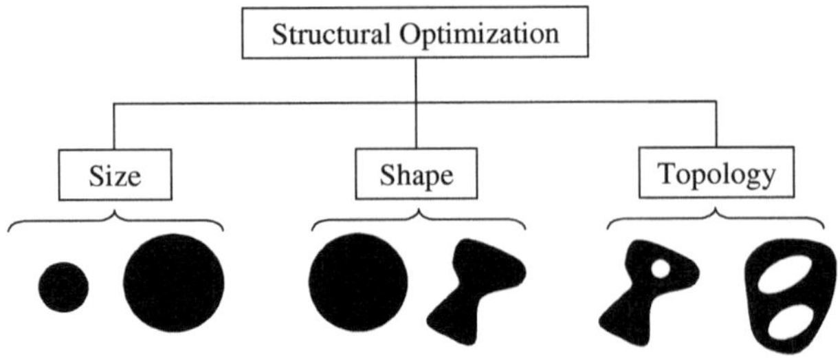

Fig. 1.4 The role of size, shape, and topology in structural optimization; ref. [10]. Reprinted from [7, Fig. 1], Copyright (2012), with permission from Inderscience Enterprises Limited

with experts in different fields, while modern software tools allow for direct import of more realistic three-dimensional (3-D) solid model computer-aided design (CAD) geometry. To assist in the process just described, the multiple physics that are commonly considered for electromechanical system applications are outlined in Chap. 2, while governing equations for such systems are reviewed in Chap. 3. Having established the proper physics and geometry, the standard finite element method (FEM) steps, per [17], then apply and include: (1) pre-processing or model setup, (2) numerically solving the problem, and (3) post-processing or evaluation of computational results.

1.4 The Role of Structural Optimization Methods

The field of structural optimization is extremely broad and generally concerned with the definition of the shape, size, and topology of a structure, as discussed in [10] and illustrated in Fig. 1.4. Broadly, the goal in a structural optimization problem is to find the optimal set of assumed design variables that maximize performance subject to predefined design constraints (e.g., weight, size, cost, etc.). Appropriately, a significant amount of research has been published encompassing topics that span a range of physical scales from the optimization of large truss structures to the optimal design of material microstructure. Efficient numerical methods for such structural optimization applications are well established and utilize both gradient-based searches [3, 4] and evolutionary algorithms [8, 10].

In this manuscript, we focus primarily on the use of gradient-based methods for the structural topology optimization of electromechanical systems, although some pertinent parametric size analysis examples are also presented. The main significance of these numerical studies is that they are representative examples of how automated methods may be exploited in arriving at optimal structural topologies for further design verification within a commercial multiphysics CAE software environment.

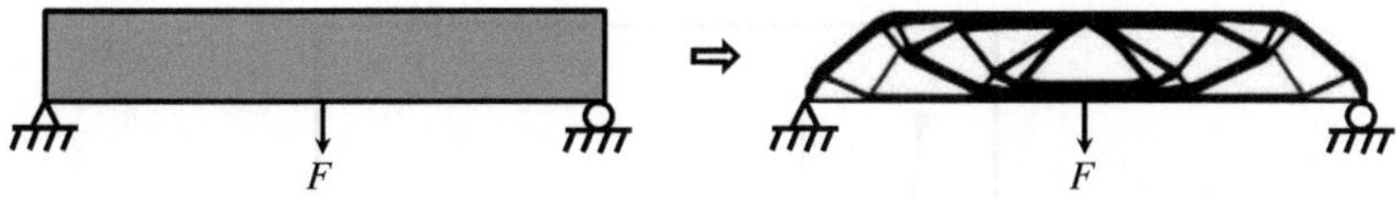

Fig. 1.5 A standard structural compliance minimization topology optimization problem is shown on the *left* for a simply supported beam. The optimal topology for minimum compliance, assuming a 50 % solid material volume fraction, is shown on the *right*

1.4.1 Topology Optimization

The origin of numerical topology optimization can be traced back to the highly influential work of Bendsøe and Kikuchi [3] published in 1988. Topology optimization has since become a rapidly developing research field, and the optimization of both single and multiphysics systems is well established for a variety of applications including actuators [26, 27], sensors [13], phononic materials [29], compliant mechanisms [32], fluid flow devices [22, 33], and fluid-structure interaction problems [34], to name a few. Given the rate at which computing power is increasing, such optimization techniques are prime for utilization in the design and development cycle of advanced electromechanical systems.

As explained in detail in [4], topology optimization is a method in which a mathematical representation of the geometry is assumed and gradient information within a finite element framework is used to guide the topological layout of a design. Discretization of a structure into many finite elements is typically required for topology optimization, and each element is assigned a design variable value, γ. For a single physics case, these design variables are often associated with a single material physical parameter. A common example is a static structural problem, and a typical objective function, F_o, is to minimize the structural compliance (i.e., deflection). In Fig. 1.5, a simply supported beam is shown with an applied load, F, in the middle. Here, the elastic modulus, E, is interpolated between $\gamma = 0$ (void) and $\gamma = 1$ (solid). Additionally, a 50 % solid material volume constraint is assigned in order to determine the optimal structural topology that minimizes deflection for a given material volume fraction; refer to the illustration on the right in Fig. 1.5. As explained in [28], a power law interpolation function, $E = \gamma^p E_o$, is commonly used to control the material state, where E_o represents the elastic modulus of a given isotropic material and p is a penalty parameter typically set to 3.

The process defined above is similar for a multiphysics problem except that additional design variables, constraints, and/or objective function terms may be considered. As an example, in a two-physics problem, the designer may define a two-term objective function that is related to two separate state variables within the design domain. Each term within this objective function may then have a weighting value associated with it, and a family of 'optimal' solutions may then be developed depending on the relative weighting or priority of each objective function term. This family

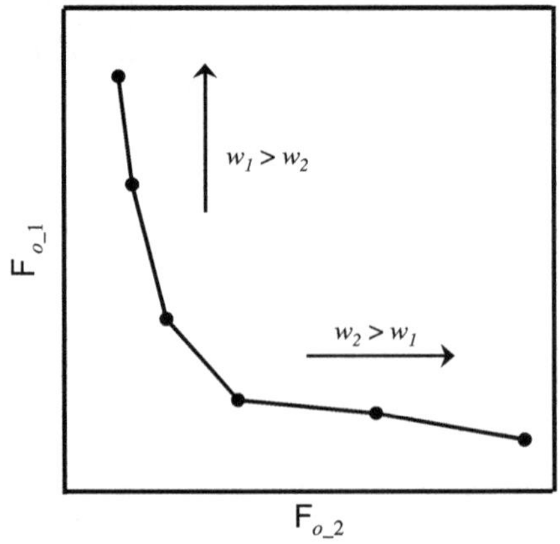

Fig. 1.6 A simple Pareto front example diagram illustrating the tradeoff between different objective function terms, F_{o_1}, F_{o_2}, in finding the optimal solution in a multi-objective optimization problem. The *vertical arrow* indicates the effect of prioritizing the weighting value, w_1, for the first objective function term, F_{o_1}, while the *horizontal arrow* shows the effect of prioritizing the weighting value, w_2, for the second objective function term, F_{o_2}

of optimal solutions may be referred to as a Pareto front as demonstrated in [20], and the general concept is shown in Fig. 1.6.

More specifics regarding the form of the non-linear optimization problem that is being solved in the process described above will be provided in Chap. 4. Discussion is provided throughout the text regarding the manual selection of, and automated methods for determining, appropriate weighting values in multiphysics, multi-objective optimization problems. Additionally, when beneficial for understanding, a flowchart of computations will be provided for the numerical examples in Chap. 5 to visualize the manner in which this optimization process ties in with the underlying multiphysics simulation. An example MATLAB®[2] optimization script is also provided in Chap. 7 to show implementation of multiple objective function terms in a multiphysics problem.

1.4.2 Size and Shape Optimization

Size and shape optimization also play a crucial role in the design of multiphysics systems. Assuming that the specific component size or shape variable (e.g., part thickness or external boundary curve/spline) that one is interested in can be adequately parametrized, many of today's advanced multiphysics finite element software programs allow for the parametric optimization of multiple variables [1, 2, 5]. Here, the

[2] MATLAB® is a registered trademark of The MathWorks, Inc.

topology (or layout) of a structure is generally known, and specific components are optimized in terms of their relative size or shape to achieve a particular objective.

A classic multiphysics example of size optimization for electromechanical systems can be found in the context of layered electronics packages, where thermally induced stresses play a key role. Thermal stress (along with vibration) is one of the leading causes of electronics failures [11]. An underlying cause of thermal stress is the mismatch in the coefficient of thermal expansion (CTE) of the various materials comprising an electronics package [31]. For example, an order of magnitude difference exists between the CTE of silicon (2.6 ppm/°C) and tin-lead solder (27 ppm/°C). Additionally, in a laminated structural assembly, the length and thickness of the various material layers is important. Thus, these geometric parameters may be efficiently optimized or 'tuned' using standard numerical tools to minimize thermal-structural effects due to CTE mismatch.

Further details regarding parametric size analysis for electromechanical systems are provided in Chap. 4, and two example studies are presented in Chap. 5. While parametric size analysis/optimization is primarily covered in this book, many of the numerical strategies may be logically extended to parametric shape optimization, as well.

References

1. Altair HyperWorks, Altair Engineering Inc. 1820 Big Beaver Rd. Troy, MI 48083
2. ANSYS Multiphysics, ANSYS Inc. Southpointe, 275 Technology Drive, Canonsburg, PA 15317
3. Bendsøe MP, Kikuchi N (1988) Generating optimal topologies in structural design using a homogenization method. Comput Method Appl M 71:197–224. doi:10.1016/0045-7825(88)90086-2
4. Bendsøe MP, Sigmund O (2003) Topology optimization: theory, methods, and applications, 2nd edn. Springer, Berlin
5. COMSOL Multiphysics, COMSOL, Inc. 1 New England Executive Park, Suite 350, Burlington, MA 01803
6. DARPA (2014) Tactical technology office: advanced vehicle make (AVM). http://www.darpa.mil/Our_Work/TTO/Programs/Adaptive_Vehicle_Make__(AVM).aspx. Accessed 12 March 2014
7. Dede EM, Lee J, Liu Y, Robert B, Yönak SH (2012) Computational methods for the optimization and design of electromechanical vehicle systems. Int J Vehicle Des 58(2–4):159–180. doi:10.1504/IJVD.2012.047383
8. Goldberg DE (1989) Genetic algorithms in search, optimization and machine learning. Addison-Wesley Publishing, Reading
9. Isight, Dassault Systèmes. 10, Rue Marcel Dassault, 78140 Vélizy-Villacoublay, France
10. Jakiela MJ, Chapman C, Duda J, Adewuya A, Saitou K (2000) Continuum structural topology design with genetic algorithms. Comput Method Appl M 186:339–356. doi:10.1016/S0045-7825(99)00390-4
11. Jiang ZQ, Huang Y, Chandra A (1997) Thermal stresses in layered electronics assemblies. J Electron Packaging 119:127–132. doi:10.1115/1.2792218
12. Kawamoto A, Matsumori T, Nomura T, Kondoh T, Yamasaki S, Nishiwaki S (2012) Topology optimization by a time-dependent diffusion equation. Int J Numer Meth Eng 93:795–817. doi:10.1002/nme.4407

13. Kim JE, Kim DS, Ma PS, Kim YY (2010) Multi-physics interpolation for the topology optimization of piezoelectric systems. Comput Method Appl M 199:3153–3168. doi:10.1016/j.cma.2010.06.021
14. Krog L, Tucker A, Rollema G (2002) Application of topology, sizing and shape optimization methods to optimal design of aircraft components. Altair Engineering, Coventry
15. Lethbridge P (2004) Multiphysics analysis. Ind Phys 12:26–29
16. Littmarck S (2001) Solving differential equations. Ind Phys 2:21–23
17. Logan DL (2002) A first course in the finite element method, 3rd edn. Brooks/Cole, Pacific Grove
18. Matsubara T, Yaguchi H, Takaoka T, Jinno K (2009) Development of new hybrid system for compact class vehicles. SAE Technical Paper 2009–01-1332. doi: 10.4271/2009-01-1332
19. MSC Software (2014) MSC Nastran. http://www.mscsoftware.com/product/msc-nastran. Accessed 12 March 2014
20. Nomura T, Sato K, Nishiwaki S, Yoshimura M (2007) Multi-disciplinary multi-objective topology optimization of electromagnetics and structural mechanics: for case of optimal dielectric resonator antenna designs. T Jpn Soc Mech Eng A 73:1111–1119. doi:10.1299/kikaia.73.1111
21. Nozawa N, Maekawa T, Nozawa S, Asakura K (2009) Development of power control unit for compact-class vehicle. SAE Int J Passeng Cars Electron Electr Syst 2:376–382. doi:10.4271/2009-01-1310
22. Olesen LH, Okkels F, Bruus H (2006) A high-level programming-language implementation of topology optimization applied to steady-state Navier-Stokes flow. Int J Numer Meth Eng 65:975–1001. doi:10.1002/nme.1468
23. Savander BR, Shankara P (2012) Propulsion system performance optimization—design by analysis. CD-adapco Dynamics 33:47–50
24. Schoofs AJG (1993) Structural optimization history and state-of-the-art. In: Dijksman JF, Nieuwstadt FTM (eds) Topics in applied mechanics. Kluwer Academic Publishers, Netherlands, pp 339–345
25. Shigley JE, Mischke CR, Budynas RG (2004) Mechanical engineering design, 7th edn. McGraw-Hill, New York
26. Sigmund O (2001) Design of multiphysics actuators using topology optimization—Part I: one-material structures. Comput Method Appl M 190:6577–6604. doi:10.1016/S0045-7825(01)00251-1
27. Sigmund O (2001) Design of multiphysics actuators using topology optimization—Part II: two-material structures. Comput Method Appl M 190:6605–6627. doi:10.1016/S0045-7825(01)00252-3
28. Sigmund O (2001) A 99 line topology optimization code written in Matlab. Struct Multidiscip O 21:120–127. doi:10.1007/s001580050176
29. Sigmund O, Jensen JS (2003) Systematic design of phononic band-gap materials and structures by topology optimization. Philos T Roy Soc A 361:1001–1019. doi:10.1098/rsta.2003.1177
30. Thilmany J (2010) Multiphysics: all at once: physical phenomena-and engineers-rarely work in isolation, so simulation software is addressing those facts. Mech Eng CIME 132:39–41
31. Timoshenko S (1925) Analysis of bi-metal thermostats. J Opt Soc Am 11:233–255. doi:10.1364/JOSA.11.000233
32. Yin L, Ananthasuresh GK (2002) A novel topology design scheme for the multi-physics problems of electro-thermally actuated compliant micromechanisms. Sensor Actuat A-Phys 97–98:599–609. doi:10.1016/S0924-4247(01)00853-6
33. Yoon GH (2012) Topological layout design of electro-fluid-thermal-compliant actuator. Comput Method Appl M 209–212:28–44. doi:10.1016/j.cma.2011.11.005
34. Yoon GH, Jensen JS, Sigmund O (2007) Topology optimization of acoustic-structure interaction problems using a mixed finite element formulation. Int J Numer Meth Eng 70:1049–1075. doi:10.1002/nme.1900
35. Zienkiewicz OC, Taylor RL, Zhu JZ (1967) The finite element method: its basis and fundamentals, 1st edn. Elsevier Butterworth-Heinemann, Burlington

Chapter 2
Overview of Physics for Electromechanical Systems

In this chapter, an overview of physics is provided for electromechanical systems concerned with electronic components, low frequency magnetic components, higher RF components, and motors and actuators. As shown in Fig. 2.1, several relevant physical scales exist when considering electromechanical systems ranging from the material or device level to the component, subsystem, or fully assembled system level. Microelectromechanical systems (MEMS), which refer to devices of size less than 1 mm but greater than 1 μm, are multiphysics by nature and fall within the broader category of electromechanical systems. However, several authoritative texts related to the design and optimization of MEMS currently exist, and the reader is referred to the literature for a comprehensive review of the topic; see for example [8, 9, 11, 20].

The majority of the numerical examples presented in this book are focused on design optimization at the component or sub-component level. Thus, in the context of these somewhat larger scales, the coupling of several physical phenomena may be defined, as shown in Fig. 2.2, where the systems considered in this text are categorized. This flowchart provides a high-level overview of interactions that engineers must consider when designing electromechanical systems. More specifics are provided on the multiple physical interactions encountered for electronic system components in Sect. 2.1. From there, the physics involved in the simulation of low frequency magnetic components including inductors and transformers are outlined in Sect. 2.2, where operating frequencies in the kHz–MHz range are generally considered. Radio frequency (RF) components operating in the MHz–GHz frequency range are subsequently described in Sect. 2.3 followed by motors and actuators in Sect. 2.4.

E. M. Dede et al., *Multiphysics Simulation*, Simulation Foundations, Methods and Applications, DOI: 10.1007/978-1-4471-5640-6_2,

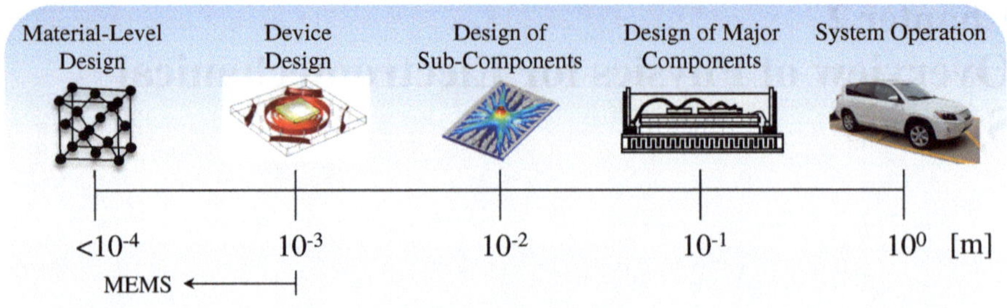

Fig. 2.1 Range of physical scales encountered in the design of electromechanical systems

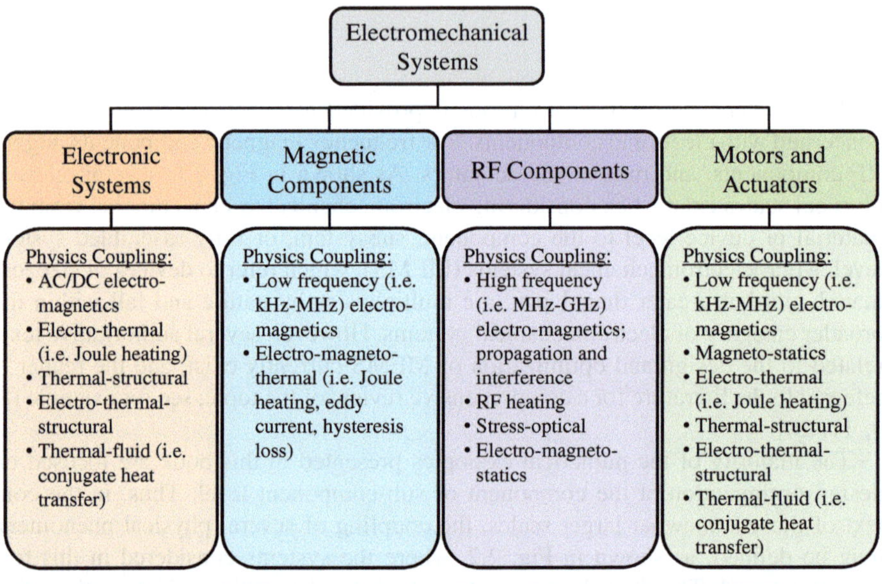

Fig. 2.2 Commonly encountered physics couplings for electromechanical system components including electronic systems, low frequency magnetics, radio frequency (RF) components, and motors and actuators

2.1 Electronic System Components

Modern electronic equipment is highly complex, and the effect of electric current flow and the resulting device power dissipation on the temperature and reliability of the package is significant. In Fig. 2.3, a representative structure is shown to illustrate the typical components found in an electronics package. While this package features a planar integrated circuit (IC) device, the main features of the structure are similar to those found in packages with discrete devices.

Generally, electric current is applied to the device, which generates an electro-magnetic field and dissipates power in the form of heat. The device is attached to a

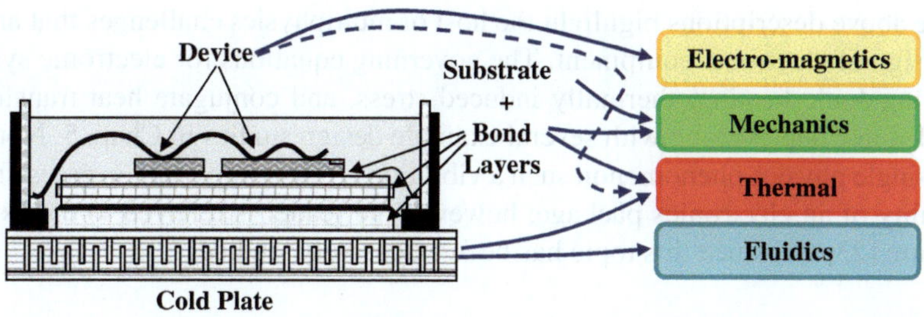

Fig. 2.3 Cross-section schematic of a representative electronics package highlighting the physics involved with various components. Although a planar type of structure is shown, the illustrated components (i.e., device, substrate, bond layer/TIM, and a cold plate/heat sink) are common to most electronics applications

substrate such as a multi-layer direct-bonded structure, or in some cases a printed circuit board (PCB), where electrical connections are made. The device is electrically and thermally coupled to the substrate via a bond layer using solder and/or some other thermal interface material (TIM). The cold plate (or heat sink) may then be attached in an electrically isolated fashion to either the bottom of the substrate or in some cases directly to the top of the device depending on the various electrical connections that are required in the final application.

Referring now to Fig. 2.2 in conjunction with Fig. 2.3, several physical interactions occur in an electronics package. These physical interactions are typically initiated by applying electrical power to a semiconductor device such as a diode or transistor. The device is usually constructed out of silicon, germanium, gallium arsenide, or an organic semiconductor, and Joule (or resistive) heating occurs due to device power dissipation. During device steady-state and transient operation, structural deformation often occurs as a result of the material coefficient of thermal expansion (CTE) mismatch between the various package layers, which may be composed of different types of metals, ceramics, plastics or other compliant TIMs. This structural deformation leads to thermally induced stress in the various package components, and while Joule heating may be examined as a standalone problem, this heating phenomenon may also be coupled into a thermal-structural multiphysics analysis.

The miniaturization of electronics enables compact form-factor designs. However, high power devices and smaller package sizes lead to higher power densities. As a result, efficient cooling of electronics is an enabling technology, and some combination of heat conduction, convection, and radiation must typically be examined in analyzing the performance of an electronics system. Conjugate (solid-fluid) heat transfer is a classic situation in which multiphysics analysis is especially important; here, heat is transported by conduction in the package plus either free or forced convection via a fluid that must efficiently interact with a heat sink. While this topic has been extensively studied, see for example [12, 14, 15, 18], the goal in addressing it here is to provide a unique set of numerical tools for the optimization of electronics cooling structures and associated fluid distribution manifolds.

The above descriptions highlight the host of multiphysics challenges that arise in the design of electronic equipment. The governing equations for electronic systems involving Joule heating, thermally induced stress, and conjugate heat transfer are provided in Chap. 3 along with several example design studies in Chap. 5. Note that other single physics phenomenon such a vibration are also important in assessing the reliability of an electronics package; however, the reader is referred to the existing literature [23, 24] since this topic has been extensively studied.

2.2 Magnetic Components

Magnetic components (e.g., inductors and transformers) are essential components in the construction of electrical circuits. Figure 2.4 shows a schematic of a representative magnetic component. Copper wires are wrapped around a core, which is generally made of an iron-related material. Magnetic components transform electric energy into magnetic energy, and vice-versa. More specifically, the electric current in the wires generates a magnetic field, which might again induce an electric current. In order to facilitate the conversion between electric and magnetic energy, the core material is inserted inside the windings of the wires because the resistance of the magnetic field (i.e., reluctance) of the core material is much lower than that of air. This electromagnetic energy conversion enables us to change the waveform of an electrical signal or the related power. During the conversion process, a portion of the total device energy is dissipated in the form of heat. Thus, a heat sink might be attached to the device to prevent overheating of the magnetic component.

Figure 2.2 together with Fig. 2.4 illustrates the various physical interactions that occur in magnetic components. Accurately modeling the coupling of electric and magnetic fields is key to the performance analysis of magnetic components. The relevant frequency range of the electromagnetic fields for the magnetic components considered herein generally spans kHz–MHz frequencies, and this frequency range is treated as a low frequency regime. The associated electromagnetic energy is dissipated as heat both in the component wires and core material. In the wires, power is dissipated due to Joule heating, which is a result of the current flow in the electrically conducting wires. In the core material, power dissipation in the form of heat occurs due to two phenomena: (1) Joule heating due to eddy currents (i.e., time variation of the magnetic field which induces current within electric conductors and associated resistive heating) and (2) hysteresis heating (i.e., energy loss dissipated in the form of heat that occurs during the repeated process of magnetization and demagnetization). The temperature rise due to heat affects the electric and magnetic material properties (e.g., electric conductivity and magnetic permeability). Thus, the coupled electro-magneto-thermal problem must be addressed in the analysis of magnetic components.

The research trend in magnetic component design is to miniaturize the component size. Particularly, magnetic components are known to be the bulkiest components in power applications. A fundamental way to reduce the size of magnetic components

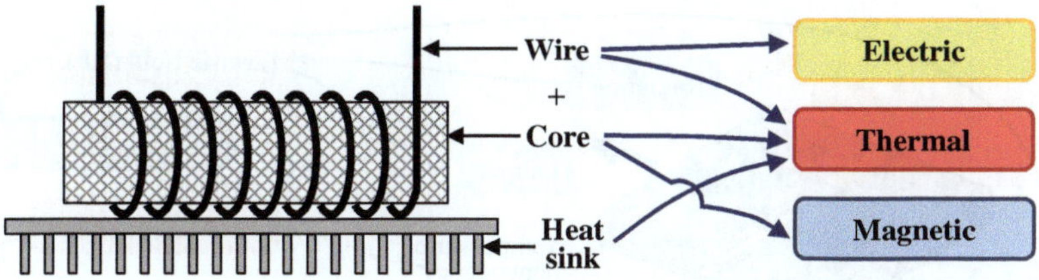

Fig. 2.4 Side view schematic of a representative low-frequency magnetic device with defined physics couplings. Although an I-shaped type of inductor structure is shown, the illustrated components (i.e., wire, core, and heat sink) are common to most magnetic components

is to increase the operating frequency. This frequency increase enables us to raise the reactive impedance without an increase in component size. However, a higher operating frequency means greater energy loss since core losses due to eddy current and hysteresis loss are proportional to the time variation of the magnetic field. The high losses of magnetic components cause not only an efficiency problem, but a thermal management problem, both of which deteriorate overall performance. To deal with these issues, multiphysics simulations considering coupled effects may be a fundamental and effective approach. The governing equations for multiphysics analysis involving low-frequency electric/magnetic fields, and Joule/hysteresis heating are provided in Chap. 3 along with several design example studies in Chap. 5. Note that basic analysis principles and design rules based on fundamental physical interpretations have been extensively investigated; for this, refer to the literature [3, 5, 7].

2.3 RF Components

Modern technologies make machines, such as computer controlled machines or unmanned vehicles, intelligent. In the commercial market, these technologies improve efficiency and safety, which are two major responsibilities of both manned and unmanned vehicle industries. Such intelligent technologies are built upon sensing and wireless communication devices, which in turn are based on electromagnetic wave devices.

Radio frequency (RF) systems represent a class of electromagnetic systems that have a high operational frequency range, and related numerical analysis schemes must be able to handle electromagnetic wave propagation at frequencies ranging from 3 kHz to 300 GHz. However, sometimes, the same numerical analysis scheme is applicable for frequencies beyond the optical range, such as infrared, visible, and ultraviolet frequency bands. Thanks to recent progress in micro or nano machining technology, controlling light waves in the RF regime via electronics has become feasible, and electromagnetic simulations at related scales have become more meaningful [13, 22].

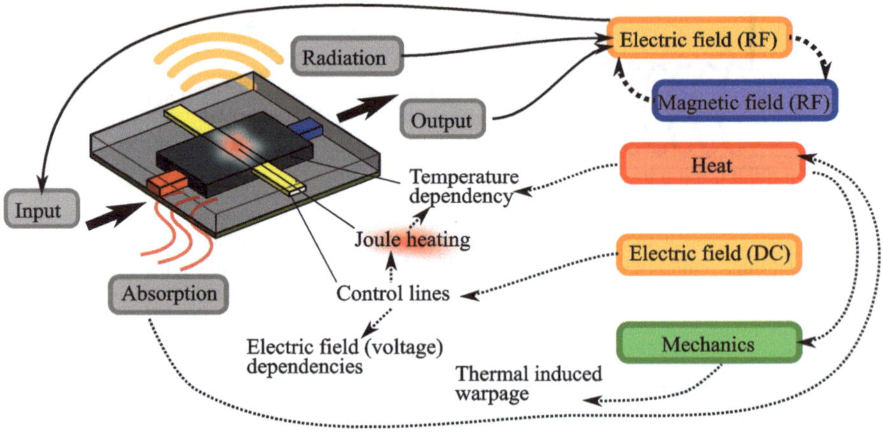

Fig. 2.5 Schematic of a representative RF device with typical multiphysics interactions

Similar to other electronic devices, one of the major multiphysics phenomenon present in RF devices is electromagnetic-thermal coupling (i.e., RF heating), as described in Fig. 2.2 and illustrated in the overview image in Fig. 2.5. Any medium, with the exception of vacuum, may contain a loss component in its material property such as the ohmic loss of a conductor and the imaginary part of the complex permittivity in dielectrics. Electromagnetic fields are absorbed by these loss components, and the related energy is dissipated in the form of heat. A typical application that exploits this phenomenon is microwave heating, and the absorption of electromagnetic waves by the human body (e.g., via cell phones) is a very important research topic with ramifications that are not yet fully understood. Here, electromagnetic analysis is heavily utilized to predict a specific absorption rate (SAR) to satisfy guideline criteria.

Additionally, most materials have temperature-dependent physical properties that form bidirectional coupling between high frequency systems and thermal transport. This coupling is very crucial in small-scale photonics systems and is often exploited for active modulation combined with Joule heating; this is often referred to as thermo-optical control. Furthermore, the thermal expansion of a material may cause structural changes of the RF device, which in terms of geometry, may give rise to changes in the electromagnetic field.

Another aspect of thermal coupling is heat radiation via infrared light. Technically, thermal radiation is an electromagnetic wave propagation effect, which peaks in the infrared range of the electromagnetic spectrum. Therefore, it is possible to control the heat radiated by a device by designing the electromagnetic behavior of the device surface structure. A typical application of this idea is the surface structural design of a thermo-photo-voltaic system [16, 26].

Beyond heat transfer, the coupling of electromagnetic fields at different frequencies with static fields is also important. It is very common to control RF or optical devices using a DC bias voltage or different operational frequencies. From a macroscopic point of view, such effects (e.g., Pockel's effect [4], Kerr effect [10]) can be described in terms of material non-linearity. Or, as previously mentioned, the use of thermo-optical coupling is a common approach to optical modulation by DC bias. Similar free carrier plasma effects [17] control the amount of free carriers on the optical path to change the permittivity by an external field. Another common coupling between electromagnetic waves and static magnetic fields is Faraday's effect. Here, the polarization direction of an electromagnetic wave rotates when the wave is exposed to a magnetic field. By combining this effect with a polarization-dependent structure, it is possible to control an electromagnetic wave via a static magnetic field, and one important related application is a ferrite circulator [6].

On the other hand, unlike other electronic systems, magnetic forces are reduced as frequency increases due to difficulties in confining the magnetic field. As scale decreases, other forces become important including Coulomb forces and optical gradient forces [25]. Coulomb forces are the forces between electric charges, and these forces are commonly exploited in MEMS devices. An optically graded force is a force generated in relation to the gradient of the electric field energy (e.g., in optical tweezers).

As electromagnetic frequency increases, the associated wave length decreases. This inverse relationship results in a reduction of physical size. Hence, small geometrical changes have a large effect on electromagnetic behavior. Accordingly, by using piezoelastic materials it is possible to create a strong coupling between strain and electromagnetic fields. One application is a surface acoustic wave (SAW) filter. Finally, in optics, there is also a coupling between refractive index, or electric permittivity, and stress, that is, photoelasticity. This effect is also utilized in acousto-optics and photoelastic modulators.

Thus, many RF device applications exist, and the reader is referred to the literature for numerous examples. In this text, governing equations for high frequency electromagnetic systems are covered in Chap. 3. Then, a RF microstrip device design example is presented in Chap. 5 followed by the multiphysics design of a dielectric resonator antenna, where the focus of the study is to optimize both electromagnetic and structural response.

2.4 Motors and Actuators

Recently, electric motors and actuators have been extensively used as fundamental components of future vehicle drive-trains and robotic systems. A possible scenario is that electric motors will eventually replace traditional internal combustion engines for future eco-friendly vehicles. Actuators are essential in the development of drive-by-wire technologies, which aim to replace traditional mechanical control systems with electronic systems. Figure 2.6 illustrates a schematic of a linear actuator structure that is presented in a simplified form for ease of understanding. The permanent magnet (PM) and electro-magnet (composed of yoke and wires) generate a magnetic field loop that passes within the plunger and through the yoke (or rotor and stator, respectively, for a motor). The generated magnetic field causes a magnetic force

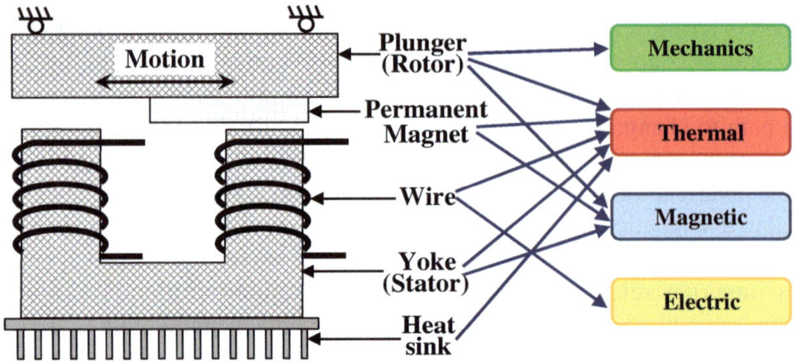

Fig. 2.6 Conceptual schematic view of a simple linear actuator. The featured components of the device (i.e., permanent magnet, wire, etc.) are common to not only actuators, but electric motors

that acts on the plunger. Consequently, the electric energy supplied to the wires is converted into mechanical energy at the plunger through the magnetic energy surrounding the device. During this energy conversion processes, thermal energy is produced due to power loss in the electrical wires and magnetic materials (i.e., yoke and plunger). Logically, a heat sink (or cooling jacket) is then attached for thermal management of the device.

The physical interactions that occur in motors and actuators are very similar to those present in magnetic components with the exception of the mechanics-related physics. Low-frequency electromagnetic field coupling is again a fundamental phenomenon that must be considered in the performance prediction of motors and actuators. The heat generation effect due to Joule and hysteresis losses (i.e., temperature rise due to heat dissipation) is identical to the effect explained in relation to magnetic components. The main difference is that motors and actuators are devices that generate force (or torque) and thus mechanical power. Therefore, it is necessary to understand the dynamics, structural, and vibration characteristics related to the plunger (or rotor) movement. Also, this mechanical behavior may have cascading effects and influence the electromagnetic characteristics. Thus, the coupled electromagnetic-thermal-mechanical problem needs to be addressed in the analysis and design of motors and actuators.

An active research area with regard to motors and actuators is the development of methods for simultaneously increasing power density (i.e., generated power amount per unit device size) and efficiency (i.e., mechanical energy output per consumed electrical energy input). The enhancement of these two performance metrics is required in order to improve the mobility of motors and actuators. Many extensive research studies have been conducted, and one technical approach is the development of new materials based on nano-technologies. Such materials hold promise to offer better material properties (e.g., permanent magnets that have higher residual magnetic flux density or iron that has higher magnetic permeability with low electric conductivity). Improving electric power conversion devices such as inverters and converters is a second potential approach to achieve high power density and efficiency, where engineers seek an optimal electrical energy input waveform. Lastly, design optimization based on multiphysics simulation, as addressed in this book, may also be an effective technique for motor and actuator performance enhancement.

Simulation methods that consider electromagnetics and force generation, relative to power density and efficiency, have been extensively studied; see for example [1, 2, 19, 21]. Thus, related governing equations for the multiphysics analysis of motors and actuators will be provided in Chap. 3 along with several design example studies in Chap. 5. However, it should be mentioned that understanding thermal and vibration performance is another critical piece of the puzzle in the analysis and design of motors and actuators. Fully coupled simulation methods that handle electromagnetics, force generation, heat transfer, and structural dynamics are not yet mainstream due to their inherent complexity, although future research into motors and actuators may address all of these effects.

References

1. Bastos JPA, Sadowski N (2003) Electromagnetic modeling by finite element methods. Marcel Dekker, New York
2. Bianchi N (2005) Electrical machine analysis using finite elements. CRC Press, Boca Raton
3. Bossche AVD, Valchev VC (2005) Inductors and transformers for power electronics. CRC Press, Boca Raton
4. Chmielak B, Waldow M, Matheisen C, Ripperda C, Bolten J, Wahlbrink T, Nagel M, Merget F, Kurz H (2011) Pockels effect based fully integrated, strained silicon electro-optic modulator. Opt Express 19:17212–17219. doi:10.1364/OE.19.017212
5. Del Vecchio RM, Poulin B, Feghali PT, Shah DM, Ahuja R (2010) Transformer design principles. CRC Press, Boca Raton
6. Fay CE, Comstock RL (1965) Operation of the ferrite junction circulator. IEEE Trans Microw Theory Tech 1:15–27. doi:10.1109/TMTT.1965.1125923
7. Flanagan WM (1992) Handbook of transformer design and applications, 2nd edn. McGraw-Hill, New York
8. Gad-el-Hak M (2002) The MEMS handbook. CRC Press, Boca Raton
9. Gianchandani Y, Tabata O, Zappe H (2008) Comprehensive microsystems, vol 1–3. Elsevier, Amsterdam
10. Hache F, Ricard D, Flytzanis C, Kreibig U (1988) The optical kerr effect in small metal particles and metal colloids: the case of gold. Appl Phys A Mater Sci Process 47:347–357. doi:10.1007/BF00615498
11. Hsu T-R (2008) MEMS and microsystems: design, manufacture, and nanoscale engineering, 2nd edn. Wiley, Hoboken
12. Incropera FP (1999) Liquid cooling of electronic devices by single-phase convection. Wiley-Interscience, New York
13. Joannopoulos JD, Meade RD, Winn JN (1995) Photonic crystals: molding the flow of light, 1st edn. Princeton University Press, New Jersey
14. Kays WM, London AL (1998) Compact heat exchangers. Krieger Publishing Company, Malabar
15. Kim S-J, Lee S-W (1996) Air cooling technology for electronic equipment. CRC Press, Boca Raton
16. Lin SY, Moreno J, Fleming J (2003) Three-dimensional photonic-crystal emitter for thermal photovoltaic power generation. Appl Phys Lett 83:380. doi:10.1063/1.1592614

17. Mikami O, Nakagome H (1984) Waveguided optical switch in InGaAs/InP using free-carrier plasma dispersion. Electron Lett 20:228–229. doi:10.1049/el:19840153
18. Remsburg R (1998) Advanced thermal design of electronic equipment. International Thomson Publishing, Florence
19. Salon SJ (1995) Finite element analysis of electrical machines. Kluwer Academic Publishers, Boston
20. Senturia SR (2001) Microsystem design. Springer, New York
21. Silvester PP, Ferrari RL (1996) Finite elements for electrical engineers, 3rd edn. Cambridge University Press, New York
22. Soref R (2006) The past, present, and future of silicon photonics. IEEE J Sel Top Quant 12:1678–1687. doi:10.1109/JSTQE.2006.883151
23. Steinberg DS (2000) Vibration analysis for electronic equipment. Wiley, New York
24. Steinberg DS (2001) Preventing thermal cycling and vibration failures in electronic equipment. Wiley, New York
25. Tlusty T, Meller A, Bar-Ziv R (1998) Optical gradient forces of strongly localized fields. Phys Rev Lett 81:1738–1741. doi:10.1103/PhysRevLett.81.1738
26. Wedlock BD (1963) Thermo-photo-voltaic energy conversion. Proc IEEE 51:694–698. doi:10.1109/PROC.1963.2261

Chapter 3
Governing Equations for Electromechanical Systems

In this chapter, the governing equations for the electromechanical systems described in Chap. 2 are explained in detail. Joule heating, thermal stress, and conjugate heat transfer problems are first introduced. Low frequency electromagnetic systems are then addressed followed by high frequency electromagnetics. Prior to providing the detailed equations for each physical system, a conventional forward analysis solution of a single physics structural mechanics example problem is provided as a point of reference for the steps involved in standard finite element analysis. For each multiphysics system that then follows, a system diagram, the governing equations, and typical system loads plus boundary conditions are provided.

3.1 Single Physics Structural Mechanics Example

The utility of the finite element method is that it may be used to produce an approximate solution to a set of equations that characterize the response of a structure by discretizing it effectively into many elements or sub-regions. This approach is common for a broad set of problems that cover a range of physics, and a representative FEA example is related to the computation of the deflection of a structure due to an applied static load.

For illustrative purposes, a two-dimensional (2-D) simply supported (i.e., transverse displacements constrained, but rotation and longitudinal displacement permitted [15]) beam comprising isotropic linear-elastic material is assumed in this introductory case, where a load is applied to the center of the structure, as shown in Fig. 3.1. Here, the reader is assumed to have some background knowledge of structural mechanics [5] and the finite element method [7, 12, 19]. As described in [12], the principle of virtual work is utilized to obtain the governing matrix system of equations for the displacement of the structure

$$\delta U = \delta W, \tag{3.1}$$

E. M. Dede et al., *Multiphysics Simulation,* Simulation Foundations, Methods and Applications, DOI: 10.1007/978-1-4471-5640-6_3,

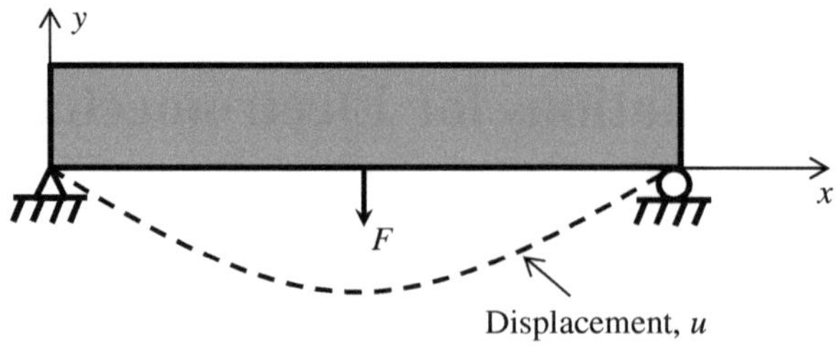

Fig. 3.1 A simply supported beam with an applied load in the center. The beam is assumed to be made of an isotropic linear-elastic material. The *dashed line* indicates the approximate deformed shape of the *lower edge* of the beam

where δU is the virtual strain energy of a finite element due to internal stress and δW is the virtual work done by the external forces on the element.

The virtual strain energy of a finite element may be written in matrix form in terms of the virtual strains, $\delta\varepsilon$, and internal stresses, σ, as

$$\delta U = \int_{v_e} \delta\varepsilon^{\mathrm{T}} \sigma \mathrm{d}v_e, \tag{3.2}$$

where v_e is the volume of the element.

Strains may be expressed in terms of the element nodal structural displacements, $\mathbf{u}$, as

$$\varepsilon = \mathbf{B}_e \mathbf{u}, \tag{3.3}$$

where $\mathbf{B}_e$ is an array of derivatives of the element 'interpolation' or nodal shape functions, which are typically linear, quadratic, or higher order polynomials that approximate the displacements of the element [7, 12]. Additionally, stresses are related to strains via the stiffness tensor, $\mathbf{C}$, using

$$\sigma = \mathbf{C}\varepsilon. \tag{3.4}$$

Substituting Eqs. (3.3) and (3.4) into Eq. (3.2) leads to

$$\delta U = \int_{v_e} \delta\mathbf{u}^{\mathrm{T}} \mathbf{B}_e{}^{\mathrm{T}} \mathbf{C} \mathbf{B}_e \mathbf{u} \mathrm{d}v_e. \tag{3.5}$$

The virtual work done by the nodal structural forces, $\mathbf{f}_S$, acting on an element may also be expressed as

$$\delta W = \delta\mathbf{u}^{\mathrm{T}} \mathbf{f}_S, \tag{3.6}$$

and equating this expression with Eq. (3.5) gives

$$\int_{v_e} \delta \mathbf{u}^{\mathrm{T}} \mathbf{B}_e{}^{\mathrm{T}} \mathbf{C} \mathbf{B}_e \mathbf{u} \mathrm{d}v_e = \delta \mathbf{u}^{\mathrm{T}} \mathbf{f}_S, \tag{3.7}$$

from which it can be deduced that

$$\int_{v_e} \mathbf{B}_e{}^{\mathrm{T}} \mathbf{C} \mathbf{B}_e \mathrm{d}v_e \mathbf{u} = \mathbf{f}_S. \tag{3.8}$$

Equation (3.8) may be rewritten in the commonly recognized form

$$\mathbf{k}_S \mathbf{u} = \mathbf{f}_S, \tag{3.9}$$

where

$$\mathbf{k}_S = \int_{v_e} \mathbf{B}_e{}^{\mathrm{T}} \mathbf{C} \mathbf{B}_e \mathrm{d}v_e, \tag{3.10}$$

is the element structural stiffness matrix, $\mathbf{u}$ is the vector of element nodal displacements, and $\mathbf{f}_S$ is the vector of applied element structural forces. Once each element stiffness matrix is computed, a standard assembly process is used to generate a global structural stiffness matrix, $\mathbf{K}_S$ [SI unit: N/m], and force vector, $\mathbf{F}_S$ [SI unit: N], to solve for the global displacements, $\mathbf{U}$ [SI unit: m], of a given structure

$$\mathbf{K}_S \mathbf{U} = \mathbf{F}_S; \tag{3.11}$$

refer to [7, 12, 19] for additional details.

The governing equations above represent an approximation to the solution of the actual structural mechanics beam deflection problem, and this approximate solution may be found using a wide range of commercially available software programs. However, regardless of the selected software, applied FEA usually involves three steps:

- *Pre-processing*: The structural geometry is defined (sometimes directly from an imported CAD file) along with material properties, structural boundary conditions (BCs), and applied loads. The computational mesh is also typically defined in this

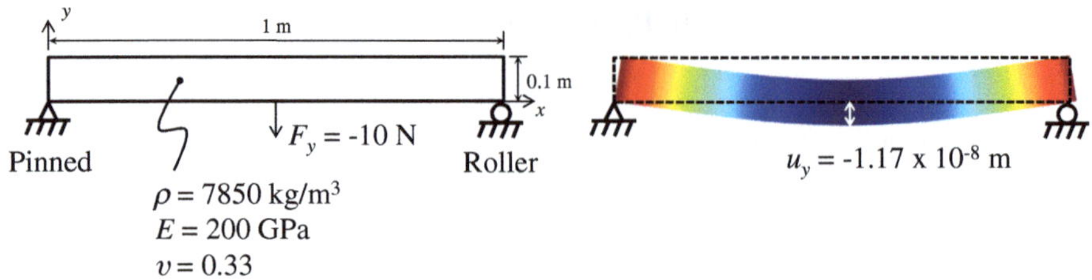

Fig. 3.2 The *left side* image illustrates the definition of the structural geometry, material physical properties, boundary conditions, and applied loads for a simply supported beam required as part of the FEA pre-processing; a 2-D plane-strain approximation is assumed for the planar model. The *right side* image shows the post-processed deformed shape of the structure with y-direction displacement contours (*blue colored contours* = larger displacement)

part of the modeling process, which along with the model BCs and loads defines a system of algebraic equations.

- *Numerical Analysis*: Numerical algorithms are used to solve the linear or non-linear algebraic system of equations defined in the preprocessing step.
- *Post-processing*: The output obtained through the prior analysis is post-processed and conveyed through the generation of data and figures that quantify the state variable results (e.g., deformed structural geometry, temperature field, and magnetic field).

For clarity, the pre and post-processing steps are illustrated in Fig. 3.2. In the image on the left, the structural geometry, material physical properties (including density, ρ, elastic modulus, E, and Poisson's ratio, ν), boundary conditions, and applied loads for a simply supported beam are provided as part of the FEA pre-processing, where a 2-D plane-strain approximation is assumed for the planar model [12]. The numerical analysis then proceeds, where a convergence criteria for the selected solution algorithm is typically satisfied. Finally, the vertical deflection and deformed shape of the structure may be obtained through post-processing of the computational results, as shown in the right side image in Fig. 3.2. For simple structural geometries, such numerical results are easily compared with standard analytical predictions [15].

Note that the sizing of elements (plus associated mesh refinement) and the treatment of concentrated point loads play an important role in obtaining an accurate numerical solution. The specific physics involved, geometry details, nature of the loads, boundary conditions, and numerical quantity of interest all affect the level of discretization used in the numerical model and method of applying the loads. Regions where large gradients occur in the state variables often require a finer mesh, and the application of concentrated loads needs to be carefully considered relative to the particular quantity of interest. In practice, however, every physical situation is different and one should be wary of broadly applying general rules. The selected modeling strategy should be based on prior experience (coursework and/or applied) or appropriate engineering references to related work.

In summary, while the principle of virtual work was exploited in developing the governing equations for the structural mechanics beam deflection problem above, other approximation methods for the derivation of the governing equations for different physical systems may logically be used. Regardless, the overall 'big-picture' process explained above may be translated to most physical scenarios including the multiphysics systems described in detail throughout the remainder of this chapter.

3.2 Joule Heating

Joule, or resistive heating in a material or device arises from power dissipated in the form of heat when an electric current is passed through a resistive medium. For electromechanical systems, the resistive medium may be an electrical conductor or an electronics device. The loads and boundary conditions for a general system subject to Joule heating are shown in Fig. 3.3. It is constructive to start with the heat diffusion (conduction) equation [8, 10], which is related to the law of conservation of energy or the First Law of Thermodynamics, to describe the associated heat transfer in an assumed domain, Ω,

$$\rho C_p \frac{\partial T}{\partial t} = \nabla \cdot (k \nabla T) + Q, \tag{3.12}$$

where the conductor density, specific heat capacity, and thermal conductivity are given as ρ [SI unit: kg/m^3], C_p [SI unit: J/(kg K)] and k [SI unit: W/(m K)], respectively, and T [SI unit: K] is the temperature state variable. For response under steady-state conditions, the term on the left hand side of Eq. (3.12) is zero.

Regarding Joule heating, the heat source term, Q, [SI unit: W/m^3] on the right hand side of this expression arises from an electromagnetic heat source

$$Q = \mathbf{J} \cdot \mathbf{E}, \tag{3.13}$$

where $\mathbf{E}$ [SI unit: V/m] is the electric field, and $\mathbf{J}$ [SI unit: A/m^2] is the current density.

More specifically, under steady-state conditions the divergence of the current density is zero

$$\nabla \cdot \mathbf{J} = 0. \tag{3.14}$$

Furthermore, Ohm's law provides for the current density in terms of the electric field as

$$\mathbf{J} = \varsigma \mathbf{E} + \mathbf{J}_e, \tag{3.15}$$

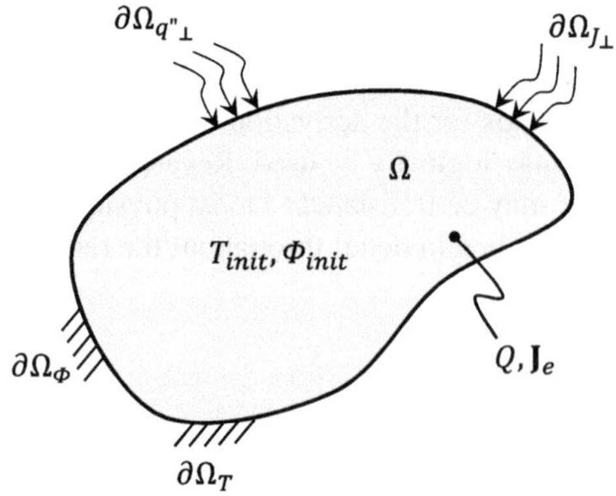

Fig. 3.3 Loads and boundary conditions for a general Joule heating problem

where ς is the electric conductivity [SI unit: S/m] of the conductive material, and $\mathbf{J}_e$ [SI unit: A/m^2] is an externally generated current density.

Finally, substituting this expression into the continuity equation and generalizing for current sources we have

$$\nabla \cdot (\varsigma \mathbf{E} + \mathbf{J}_e) = Q_j, \tag{3.16}$$

where the electric field strength may be expressed in terms of the gradient of the electric scalar potential, Φ [SI unit: V], as $\mathbf{E} = -\nabla\Phi$. The heat source due to Joule heating, Q_j, determined via Eq. (3.16) is substituted into Eq. (3.12) as the thermal load to the system.

Following the comprehensive review of coupled electrothermal effects for related microactuators [1] and MEMS [2], the solution of the electrothermal problem above requires the specification of boundary conditions in addition to an initial temperature, T_{init}, and electric potential field, Φ_{init}. As shown in Fig. 3.3, either assumed temperature, $\partial\Omega_T$, and electric potential, $\partial\Omega_\Phi$, (i.e., *Dirichlet*) boundary conditions, or assumed normal heat flux, $\partial\Omega_{q''_\perp}$, and electric current density, $\partial\Omega_{J_\perp}$, (i.e., *Neumann*) boundary conditions are applied. In the special case of a convection heat transfer boundary condition, the normal heat flux, $q''_\perp = h(T - T_o)$, is proportional to the temperature difference between the surface and the reference temperature, T_o, of an assumed surrounding fluid, where h [SI unit: W/(m^2 K)] is the assumed surface convection coefficient [8]. Although not specifically addressed here, the reader is referred to the literature for a discussion of radiation heat transfer boundary conditions [2, 8, 10].

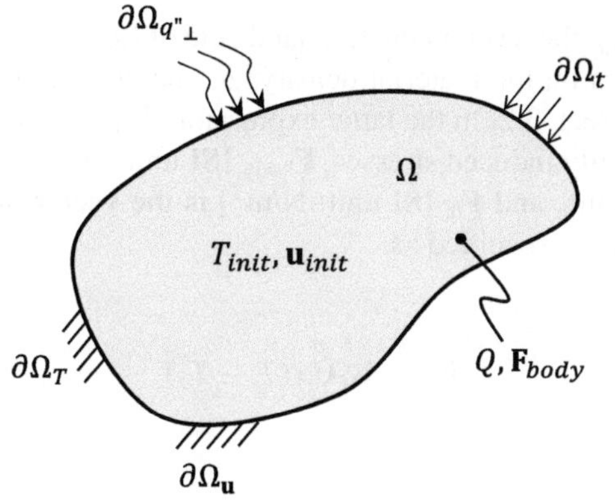

Fig. 3.4 Loads and boundary conditions for a general thermal stress problem

3.3 Thermal Stress

The topic of thermal stress is broad, and in the context of electromechanical systems there are many applications that could be considered. In particular, the subject of thermal stress and strain in electronics packaging may be broken down into several subcategories including the thermoelastic analysis of multilayered devices, determination of thermal stresses in thin films, evaluation of viscoplastic thermal damage (i.e., cumulative life-cycle) effects in die attachment materials, and the study of the thermal-mechanical loading of wirebond structures, to name a few [11]. However, to narrow the field, the topic of thermal stress in this book is focused on the fundamental case of the response of a thin laminated (i.e., multilayered) composite structure composed of non-temperature-dependent isotropic linear-elastic materials. The reader is referred to the literature for more complex topics that then build off of this essential example.

Considering the above caveats, the loads and boundary conditions for a representative system are illustrated in Fig. 3.4, and the governing equations for quasi-static thermoelasticity in a given domain, Ω, follow from [11]. These equations include the heat conduction equation and the equation of motion for the system, which may, respectively, be written as

$$\rho C_p \frac{\partial T}{\partial t} = \nabla \cdot (k \nabla T) + Q, \quad \text{and} \tag{3.17}$$

$$-\nabla \cdot \sigma = \mathbf{F}_{\text{body}} + \mathbf{F}_{\alpha}. \tag{3.18}$$

In Eq. (3.17), T [SI unit: K] is the temperature state variable, k [SI unit: W/(m K)] is the material thermal conductivity, and Q [SI unit: W/m^3] is the volumetric power density. Additionally, the term on the left hand side of Eq. (3.17) is the time transient term, which depends on the material density, ρ, and specific heat capacity, C_p, as discussed above in Sect. 3.2. In the latter expression, Eq. (3.18), σ [SI unit: N/m^2] is the matrix of thermally induced stresses, $\mathbf{F}_{\text{body}}$ [SI unit: N/m^3] is the vector of body forces per unit volume, and $\mathbf{F}_\alpha$ [SI unit: N/m^3] is the vector of thermally induced loads (per unit volume) specified as

$$\mathbf{F}_\alpha = \mathbf{B}_e \mathbf{C} \alpha (T - T_o), \tag{3.19}$$

where $\mathbf{B}_e$ is an array of derivatives of the element nodal shape functions, $\mathbf{C}$ [SI unit: N/m^2] is the stiffness tensor (refer to the single physics structural mechanics example at the beginning of this chapter), and α [SI unit: K^{-1}] is the material CTE. In addition, T_o in Eq. (3.19) is a reference temperature required for the computation of the induced thermal loads.

For steady-state analysis of thermoelastic systems, it is common to assume some constant applied volumetric power density, Q, as a thermal load to the system described by Fig. 3.4. Alternatively, a fixed temperature gradient across predetermined system boundaries may be applied. In either case, a fixed temperature, heat flux, or convection heat transfer boundary condition is then assumed, as described above in Sect. 3.2, for proper description of the heat transfer portion of the multiphysics problem. Then, following the thermal and electrical problems introduced thus far, the structural response and stress state due to the induced thermal strains and any body forces, $\mathbf{F}_{\text{body}}$, are determined through the application of appropriate displacement constraints, $\partial \Omega_{\mathbf{u}}$, or traction conditions, $\partial \Omega_t$, on the external boundaries of the domain; refer to Fig. 3.4.

3.4 Conjugate Heat Transfer

Similarly to the case of thermal stress, fluid mechanics in conjunction with thermal transport is a very broad topic with many areas of subspecialty and numerous authoritative texts; see, for example [6, 8, 14]. Here, the topic of conjugate heat transfer is limited to the case of conduction plus convection heat transfer (i.e., radiation effects are ignored), where all flows are considered to be viscous, incompressible, and generally laminar. However, notwithstanding these conditions, some discussion of conjugate heat transfer under turbulent flow will also be provided in relation to logical extensions of the design studies presented herein.

Typical loads and boundary conditions for a domain, Ω, are shown in Fig. 3.5, where the domain consists of a moving fluid region, Ω_f, and a solid body, Ω_s [14].

For steady-state fluid flows, the continuity equation (or incompressibility condition) is a mathematical statement of the principle of conservation of mass

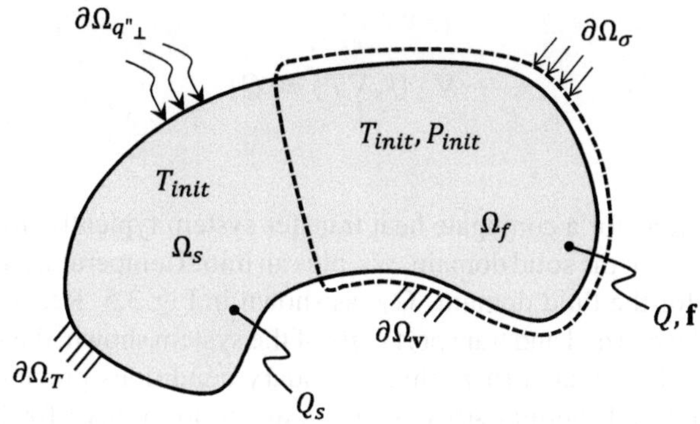

Fig. 3.5 Loads and boundary conditions for a general conjugate heat transfer problem

$$\nabla \cdot \mathbf{v} = 0, \tag{3.20}$$

where $\mathbf{v}$ is the fluid velocity vector [SI unit: m/s].

Newton's Second Law of Motion, a statement of the principle of conservation of linear momentum, is then expressed in equation form for steady-state flows

$$\rho\,(\mathbf{v} \cdot \nabla\mathbf{v}) = -\nabla P + \nabla \cdot \left\{\eta\left[\nabla\mathbf{v} + (\nabla\mathbf{v})^{\mathrm{T}}\right]\right\} + \rho\mathbf{f}, \tag{3.21}$$

where ρ is the fluid density [SI unit: kg/m^3], η is the fluid dynamic viscosity [SI unit: Pa s], P is the pressure state variable [SI unit: Pa], and $\mathbf{f}$ is the body force vector per unit mass [SI unit: N/kg].

Finally, the First Law of Thermodynamics introduced in Sect. 3.2 above is used to rewrite the steady-state heat equation with consideration of convective heat transfer

$$\rho C_p\,(\mathbf{v} \cdot \nabla T) = \nabla \cdot (k\nabla T) + Q, \tag{3.22}$$

where C_p represents the heat capacity [SI unit: J/(kg K)], k is the thermal conductivity of the fluid [SI unit: W/(m K)], Q is the volumetric power density [SI unit: W/m^3], and T is the temperature state variable [SI unit: K]. Note that in the case of the solid domain, Ω_s, in Fig. 3.5, the fluid velocity vector $\mathbf{v}$ is zero and the left hand side term in Eq. (3.22) disappears reducing the equation to the form

$$-\nabla \cdot (k_s \nabla T) = Q_s. \tag{3.23}$$

Initial conditions for a conjugate heat transfer system typically involve an initial temperature, T_{init}, for the solid domain, Ω_s, plus an initial temperature and hydrostatic pressure, P_{init}, for the fluid domain, Ω_f, as shown in Fig. 3.5. Regarding boundary conditions for the thermal and transport parts of the system shown, the same Dirichlet (temperature) and Neumann (heat flux) boundary conditions previously discussed for Joule heating and thermal stress systems again apply here for both fluid and solid portions of the domain. Also, convection and radiation conditions, as described in [14], may be applied. For the fluid dynamic portion of the system illustrated in Fig. 3.5, either the velocity components, $\partial\Omega_{\mathbf{v}}$, (i.e., *Dirichlet boundary conditions*) or total surface stress, $\partial\Omega_\sigma$, (i.e., *Neumann boundary conditions*) must be specified on the boundary of the fluid region; refer to [14] for additional explanation of these and other specialized (e.g., periodic, slip, and interfacial stress) boundary conditions.

3.5 Low Frequency Electromagnetics

This section explains three types of analysis methods for low frequency electromagnetic systems based on the finite element method [4, 16, 17]. Here, the term low frequency refers to the case in which the frequency of the electromagnetic field is lower than several hundred kilohertz. In this low frequency situation, the wavelength of the electromagnetic field is much longer than the device size, and thus the wave characteristics of electromagnetic field becomes insignificant. The original governing equations for electromagnetics, i.e., Maxwell's equations can be written as

$$\nabla \times \mathbf{E} = -\frac{\partial \mathbf{B}}{\partial t} \quad \text{Faraday's law} \tag{3.24}$$

$$\nabla \times \mathbf{H} = \frac{\partial \mathbf{D}}{\partial t} + \mathbf{J} \quad \text{Ampère's law} \tag{3.25}$$

$$\nabla \cdot \mathbf{D} = \rho_q \quad \text{Gauss's law} \tag{3.26}$$

$$\nabla \cdot \mathbf{B} = 0 \quad \text{Gauss's law of magnetics,} \tag{3.27}$$

and the constitutive laws,

$$\mathbf{D} = \epsilon \mathbf{E} \tag{3.28}$$

$$\mathbf{B} = \mu \mathbf{H} \tag{3.29}$$

$$\mathbf{J} = \varsigma \mathbf{E}, \tag{3.30}$$

where **E**, **H**, **B**, **D**, **J** and ρ_q are the electric field [SI unit: V/m], the magnetic field [SI unit: A/m], the magnetic flux density [SI unit: T], the electric displacement [SI unit: C/m^2], the electric current density [SI unit: A/m^2], and the electric charge density [SI unit: C/m^3], respectively. Also, ϵ, μ, and ς are, respectively, the electric permittivity [SI unit: F/m], magnetic permeability [SI unit: H/m], and electric conductivity [SI unit: S/m].

In the low frequency regime, the time dependent terms of the above Maxwell's equations are of little importance, and thus they may be ignored for simplification depending on the situation and application. Based on the level of simplification, it is possible to implement one of the three different analysis approaches for low frequency applications.

The first analysis method for low frequency applications is electric and magnetic field analysis. In this analysis method, the electric and magnetic fields are, respectively, derived using the electric scalar potential, Φ, and the magnetic vector potential, **A**, from Eqs. (3.24), (3.26) and (3.27) without ignoring any time-dependent terms of the original Maxwell's equation

$$\mathbf{E} = -\nabla \Phi - \frac{\partial \mathbf{A}}{\partial t}, \tag{3.31}$$

$$\mathbf{B} = \nabla \times \mathbf{A}. \tag{3.32}$$

Then, both electric and magnetic potentials are obtained by solving Eq. (3.25) with the constitutive laws, Eqs. (3.28)–(3.30). The boundary conditions for the electric and magnetic field analysis are illustrated in Fig. 3.6. To understand the boundary conditions, the analysis of a 3-D inductor can be considered. The voltage difference at the terminals of the copper winding is modeled as the Dirichlet boundary condition of electric potentials, Φ. For the insulation of the magnetic field, the zero Dirichlet condition of the magnetic potential, **A**, is applied at the boundary of the analysis domain.

This first analysis method does not simplify the original Maxwell's equations and finds both electric and magnetic fields. It is useful to calculate both the current density, **J**, and magnetic flux density, **B**, when the terminal voltages are specified in the magnetic component. However, this generally requires three dimensional analysis in order to obtain a meaningful result, which is computationally expensive. Thus, this high computational cost may limit the use of this analysis method.

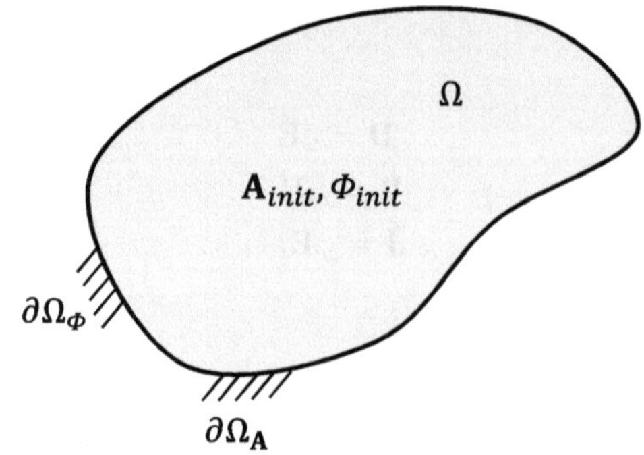

Fig. 3.6 Boundary conditions for electric and magnetic field analysis

The second computational method for low frequency electromagnetic systems uses eddy current analysis. This analysis method is useful when the eddy current plus magnetic field is of main interest, but the current density distribution due to external voltage differences is inconsequential. Here, the divergence-free magnetic flux density, **B**, is represented as the curl of the magnetic vector potential, **A**, which may be written as Eq. (3.32). The total current density in the right-hand side term of Eq. (3.25) is derived as the sum of the external current density, $\mathbf{J}_e$, and eddy current density, $\mathbf{J}_{\text{eddy}}$. Here, the external current density, $\mathbf{J}_e$, is given as an input value without considering an electric field analysis, and the eddy current density, $\mathbf{J}_{\text{eddy}}$, is calculated from the time variation of the magnetic field. Then, the total current density $\mathbf{J}_{\text{tot}}$ can be derived as

$$\mathbf{J}_{\text{tot}} = \mathbf{J}_e + \varsigma \frac{\partial \mathbf{A}}{\partial t}. \tag{3.33}$$

From Eq. (3.25) with Eqs. (3.31) and (3.33), the equation for the time-harmonic magnetic vector potential, **A**, may be derived as

$$\nabla \times \left(\frac{1}{\mu} \nabla \times \mathbf{A} \right) + j\omega\varsigma\mathbf{A} = \mathbf{J}_e, \tag{3.34}$$

where ω is the frequency of the time-harmonic vector potential, **A**. The above time-harmonic equation can be solved using the finite element method. In the 2-D case, the finite element formulation of Eq. (3.34) can be represented in the matrix form as

$$\mathbf{k}_M \mathbf{A} = \mathbf{f}_M. \tag{3.35}$$

The element magnetic stiffness matrix, $\mathbf{k}_M$, and the magnetic force vector, $\mathbf{f}_M$, are, respectively, derived as

$$k_{M,ij} = -\int_x\int_y \frac{1}{\mu}\left(\frac{\partial N_i}{\partial x}\frac{\partial N_j}{\partial x} + \frac{\partial N_i}{\partial y}\frac{\partial N_j}{\partial y}\right)\mathrm{d}y\mathrm{d}x - j\omega\int_x\int_y (\varsigma N_i N_j)\mathrm{d}y\mathrm{d}x, \tag{3.36}$$

$$f_{M,i} = \int_x\int_y (J_{e,z}N_i)\mathrm{d}y\mathrm{d}x, \tag{3.37}$$

where N_i is the shape function of i-th finite element node.

In global form after element assembly, the governing matrix equation is expressed as

$$\mathbf{K}_M\mathbf{A} = \mathbf{F}_M, \tag{3.38}$$

and by solving this equation with Eq. (3.32), the time-harmonic magnetic flux density, $\mathbf{B}$, is obtained for the whole analysis domain.

The third and final analysis method for low frequency applications is magnetostatic analysis. Using this method, all time varying terms in the governing equations are ignored by making use of the assumption of a static or quasi-static field, and the magnetic field is calculated when the external current density, $\mathbf{J}_e$ is given and/or the permanent magnet generates the residual magnetic flux density, $\mathbf{B}_r$. From Eqs. (3.25) and (3.32), the equation for the static magnetic vector potential, $\mathbf{A}$, can be derived as

$$\nabla \times \left(\frac{1}{\mu}\nabla \times \mathbf{A}\right) = \mathbf{J}_e + \nabla \times \left(\frac{1}{\mu}\mathbf{B}_r\right). \tag{3.39}$$

The above static equation may be solved again using the finite element method. In the 2-D case, the element stiffness matrix, $\mathbf{k}_M$, and the force vector, $\mathbf{f}_M$, in the finite element formulation can, respectively, be derived as

$$k_{M,ij} = -\int_x\int_y \frac{1}{\mu}\left(\frac{\partial N_i}{\partial x}\frac{\partial N_j}{\partial x} + \frac{\partial N_i}{\partial y}\frac{\partial N_j}{\partial y}\right)\mathrm{d}y\mathrm{d}x, \tag{3.40}$$

$$f_{M,i} = \int_x\int_y (J_{e,z}N_i)\mathrm{d}y\mathrm{d}x - \int_x\int_y \frac{\partial}{\partial x}\left(\frac{1}{\mu}B_{r,y}\right)\mathrm{d}y\mathrm{d}x + \int_x\int_y \frac{\partial}{\partial y}\left(\frac{1}{\mu}B_{r,x}\right)\mathrm{d}y\mathrm{d}x. \tag{3.41}$$

The matrix equation with Eqs. (3.40) and (3.41) gives us the static distribution of the magnetic flux density, $\mathbf{B}$, for the whole analysis domain.

Note that the loads and boundary conditions for both the second and third analysis methods presented in this section are illustrated in Fig. 3.7. The external current density, $\mathbf{J}_e$, is applied at the winding areas as the given load. Then, the magnetic field distribution generated by the given external current density is calculated considering eddy currents or with static conditions. For the insulation of the magnetic field, the zero Dirichlet condition is applied at the boundary of the whole analysis domain.

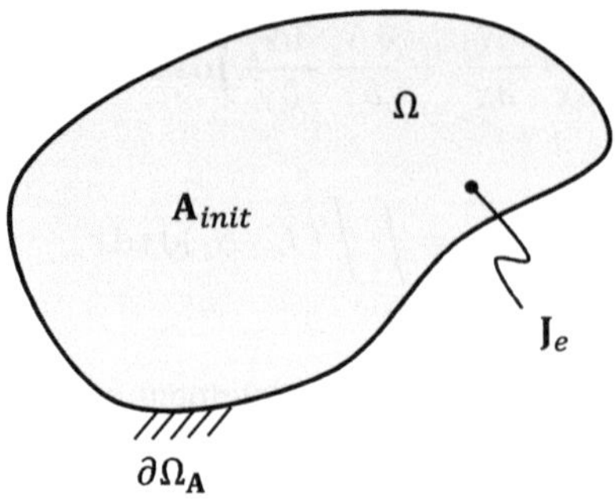

Fig. 3.7 Loads and boundary conditions for either the eddy current or magnetostatic analysis method

3.6 High Frequency Electromagnetics

In this section, two popular numerical methods are briefly described. One is the Finite Difference-Time Domain method, FDTD [18], and the other is the FEM [9]. The FDTD method is a very popular method for time transient analysis, while the FEM is usually combined with the frequency domain formulation for high frequency electromagnetic simulations.

As mentioned in the previous section, electromagnetic systems are described with Maxwell's equations

$$\nabla \times \mathbf{E} = -\frac{\partial \mathbf{B}}{\partial t} \quad \text{Faraday's law} \tag{3.42}$$

$$\nabla \times \mathbf{H} = \frac{\partial \mathbf{D}}{\partial t} + \mathbf{J} \quad \text{Ampère's law} \tag{3.43}$$

$$\nabla \cdot \mathbf{D} = \rho_q \quad \text{Gauss's law} \tag{3.44}$$

$$\nabla \cdot \mathbf{B} = 0 \quad \text{Gauss's law of magnetics,} \tag{3.45}$$

and the constitutive laws,

$$\mathbf{D} = \epsilon \mathbf{E} \tag{3.46}$$

$$\mathbf{B} = \mu \mathbf{H} \tag{3.47}$$

$$\mathbf{J} = \varsigma \mathbf{E}, \tag{3.48}$$

where **E**, **H**, **B**, **D**, **J** and ρ_q are the electric field [SI unit: V/m], the magnetic field [SI unit: A/m], the magnetic flux density [SI unit: T], the electric displacement [SI unit: C/m^2], the electric current density [SI unit: A/m^2], and the electric charge density [SI unit: C/m^3], respectively. Also, ϵ, μ and ς are, respectively, the electric permittivity [SI unit: F/m], magnetic permeability [SI unit: H/m], and electric conductivity [SI unit: S/m].

In the high frequency regime, the time dependent part of the above formulation dominates; that is, Faraday's law, Eq. (3.42), and Ampère's law, Eq. (3.43). With the Finite Difference-Time Domain method, these two equations are directly discretized with finite differences.

With Eqs. (3.42) and (3.43) and the constitutive laws, we obtain

$$\nabla \times \mathbf{E} = -\mu \frac{\partial \mathbf{H}}{\partial t} \tag{3.49}$$

$$\nabla \times \mathbf{H} = \epsilon \frac{\partial \mathbf{E}}{\partial t} + \varsigma \mathbf{E}, \tag{3.50}$$

and the FDTD discretization yields the following equations for the field in x-direction

$$E_x|^{n+1}_{i,j,k} = E_x|^{n}_{i,j,k} + \frac{\Delta t}{\epsilon} \; E_x|^{n+1}_{i,j,k} \tag{3.51}$$

$$= E_x|^{n}_{i,j,k} + \frac{\Delta t}{\epsilon} \left(\frac{H_y|^{n+1/2}_{i,j+1/2,k} - H_y|^{n+1/2}_{i,j-1,k}}{\Delta z} \right. \tag{3.52}$$

$$\left. - \frac{H_z|^{n+1/2}_{i,j,k+1/2} - H_z|^{n+1/2}_{i,j,k-1/2}}{\Delta y} \right) \tag{3.53}$$

and

$$H_x|^{n+1/2}_{i,j,k} = H_x|^{n-1/2}_{i,j,k} - \frac{\Delta t}{\mu} \; H_x|^{n+1/2}_{i,j,k} \tag{3.54}$$

$$= H_x|^{n-1/2}_{i,j,k} - \frac{\Delta t}{\mu} \left(\frac{E_y|^{n}_{i,j+1/2,k} - E_y|^{n}_{i,j-1/2,k}}{\Delta z} \right. \tag{3.55}$$

$$\left. - \frac{E_z|^{n}_{i,j,k+1/2} - E_z|^{n}_{i,j,k-1/2}}{\Delta y} \right), \tag{3.56}$$

where superscripts and subscripts on the right side of the vertical bars (|) are, respectively, temporal indices and spatial indices. FDTD uses a special finite difference grid designated as a Yee cell [18], as shown in Fig. 3.8. The magnetic field grid is offset by a half grid size from the electric field grid. The 1/2 term in the spatial indices indicates that the value is on the half offset grid. These time dependent equations are updated with an explicit method called a leap-frog algorithm, which updates one

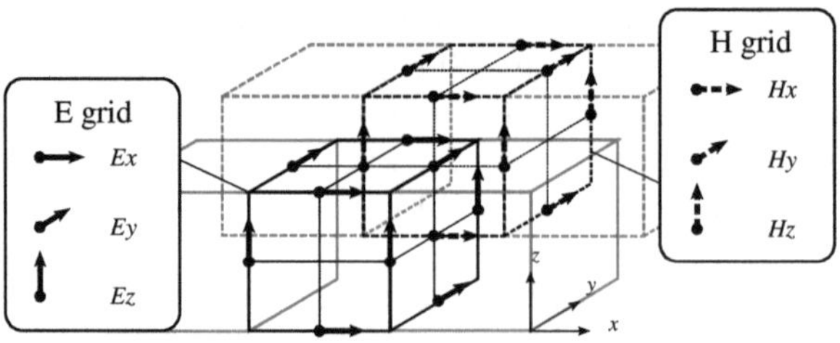

Fig. 3.8 Schematic of a Yee cell. The *solid lines* denote a cell on an electric field grid and the *dashed lines* indicate a cell on a magnetic field grid

field (the electric field) and the other field (the magnetic field) alternatively, and $1/2$ in the temporal indices represents the temporal offset of the algorithm. In a similar manner, the equations for the y and z-directions are also derived.

In the finite element method, there are several options in terms of the main state variables to be solved. For example, by following the manner in the magnetic case, it is possible to use the combination of the magnetic vector potential, $\mathbf{A}$, and scalar electric potential, Φ. Another popular option is to use either the electric field or the magnetic field. By substitution of Eqs. (3.43), (3.46) (3.47), and (3.48) into Eq. (3.42), the vectorial wave equation in the frequency domain, that is the vectorial Helmholtz equation, is derived.

$$\nabla \times \left(\frac{1}{\mu}\nabla \times \mathbf{E}\right) - \omega^2 \epsilon \mathbf{E} = -j\omega \mathbf{J} \tag{3.57}$$

$$\nabla \times \left(\frac{1}{\epsilon}\nabla \times \mathbf{H}\right) - \omega^2 \mu \mathbf{H} = \nabla \times \left(\tfrac{1}{\epsilon}\mathbf{J}\right), \tag{3.58}$$

where ω is the angular frequency of interest.

For a three-dimensional analysis, one of the above equations is discretized with Nédélec, H_{curl}, conformal elements [13], which hold tangential components of vector unknowns on the edge of the elements.

For the two-dimensional case, further simplification is applied. By choosing the out-of-plane component of the field as the unknown variable, the system reduces into a scalar Helmholtz equation. Assuming that the analysis domain resides in the x-y plane, for modes in which the electric field is polarized in the z-direction ($\mathbf{E} = \{0, 0, Ez\}^{\mathrm{T}} = E_z$), the governing equation can be written as

$$\nabla^2 E_z + k_o^2 \epsilon_r E_z = jk_o Z_o \varsigma E_z, \quad \text{where} \tag{3.59}$$

$$k_o = \omega\sqrt{\mu_o \epsilon_o}, \quad \text{and} \tag{3.60}$$

$$Z_o = \sqrt{\frac{\mu_o}{\epsilon_o}}. \tag{3.61}$$

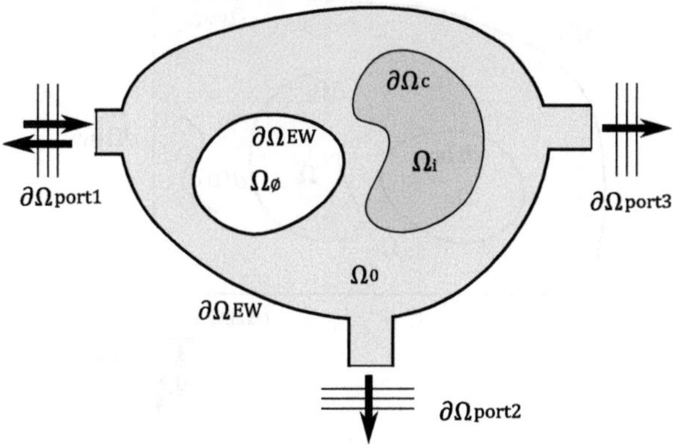

Fig. 3.9 Analysis settings for a closed high frequency electromagnetic system similar to the inside of a metallic cavity (such as a microwave oven) or a metallic waveguide

In this case, ordinary Lagrangian isoparametric elements can be used for discretization.

Regarding boundary conditions for high frequency electromagnetic systems, Fig. 3.9 shows the analysis settings for a closed region like the inside of a metallic cavity (such as a microwave oven) or a metallic waveguide. In such cases, the analysis domain is a cavity domain inside of a metal enclosure, Ω_0. The analysis domain is usually filled with air or vacuum, but it can be any dielectric or magnetic material. The boundary of the metallic cavity is truncated with a boundary, $\partial\Omega_{EW}$. The boundary condition on $\partial\Omega_{EW}$ can be a perfect electric conductor (PEC) boundary condition or an imperfect electric conductor boundary condition, namely an impedance boundary condition. The PEC boundary is sometimes referred to as an electric wall, and similarly, a perfect magnetic conductor boundary is referred to as a magnetic wall. The mathematical treatment of these boundary conditions depends on the formulation of the analysis domain. When the magnetic field is selected as the state variable, the PEC boundary condition is naturally fulfilled as a natural boundary condition. On the other hand, when the electric field is selected for the state variable, the PEC is enforced in a Dirichlet manner, such as $\nabla \times \mathbf{E} = 0$. Mathematically, the impedance boundary condition is a third type of boundary condition, i.e., a Robin boundary condition, and the term impedance boundary condition sometimes stands for the Robin boundary condition itself.

Except for the analysis model for an eigenvalue problem, there usually are some ports on the outermost boundaries of the region. These ports are special boundaries, where energy comes in and goes out of the system. Usually, at least one port is assigned to be an input port, $\partial\Omega_{\text{port1}}$, and the remaining ports are outputs, $\partial\Omega_{\text{port2}}$ and $\partial\Omega_{\text{port3}}$. The port boundary condition is also categorized as a Robin boundary condition, as $\frac{\partial\psi}{\partial\hat{\mathbf{n}}} + \Upsilon(\psi) = \upsilon$, where ψ is either the electric or magnetic field, Υ is a boundary operator, which depends on the boundary mode, and υ is the input

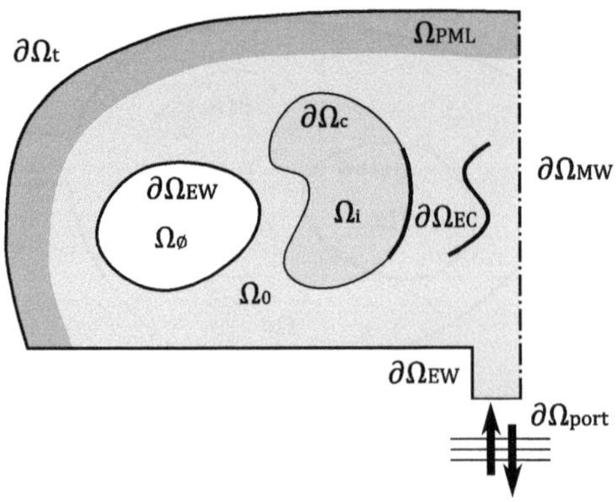

Fig. 3.10 Analysis settings for a high frequency electromagnetic system with an open boundary

mode. The boundary mode can be analytically derived or numerically calculated with eigenmode analysis of the port setting [9].

Inside the cavity in Fig. 3.9, there can be some inclusions made out of metal, dielectric or magnetic materials. For dielectric or magnetic inclusions, Ω_i, the boundary of the inclusion, $\partial\Omega_c$, is assigned an ordinary continuity condition for isotropic materials. In the case of a metal inclusion, usually, the inclusion domain is excluded from the analysis domain, i.e., $\Omega_\emptyset$. Therefore, the boundary condition surrounding $\Omega_\emptyset$ is assumed to be the same as the outermost boundaries, namely $\partial\Omega_{EW}$.

Figure 3.10 shows the analysis settings for a domain with an open boundary, $\partial\Omega_t$, with a formulation that uses the electric field as the state variable. The bottom of the domain is truncated by an electric wall, $\partial\Omega_{EW}$, and the right side is truncated by a magnetic wall. These settings represent, respectively, a metallic ground plane at the bottom and a symmetry condition, $\partial\Omega_{MW}$.

The inclusions shown in Fig. 3.10 are almost identical to those in the previous closed domain case except for the two boundaries, $\partial\Omega_{EC}$, indicated by thicker lines in Fig. 3.10. These boundaries can be a PEC or transition boundary condition, TBC, for a thin imperfect conductive sheet. One of the boundaries is a part of Ω_i, and the other is independent. The former represents that part of the inclusion boundary that is conductive, and the latter indicates that there exists a standalone conductive thin object.

The typical technique used for an open boundary is to use an absorbing boundary condition. One of the best performing absorbing boundary settings is to use a perfectly matched layer, PML [3]. A PML is not a boundary condition, *per se*, but rather represents layers of domain governed with modified Maxwell's equations to have minimum reflection on each layer boundary, while absorbing energy. If the PML works perfectly, $\partial\Omega_t$ can assume any boundary condition, e.g., PEC. However, in order to minimize the error, absorbing boundary conditions, ABCs, can be combined with a PML, per [9]. In some cases, a PML can be omitted for computational cost reduction, and a Sommerfeld radiation boundary condition is used for $\partial\Omega_t$.

References

1. Ananthasuresh GK (2008) Systematic synthesis methods. In: Gianchandani Y, Tabata O, Zappe H (eds) Comprehensive microsystems, vol 1. Elsevier, Amsterdam, pp 559–583
2. Bechtold T, Rudnyi EB, Korvink JG (2007) Fast simulation of electro-thermal MEMS. Springer-Verlag, Berlin
3. Berenger J (1994) A perfectly matched layer for the absorption of electromagnetic waves. J Comput Phys 114:185–200. doi:10.1006/jcph.1994.1159
4. Bastos JPA, Sadowski N (2003) Electromagnetic modeling by finite element methods. Marcel Dekker, New York
5. Beer FP, Johnston ER, DeWolf JT (2002) Mechanics of materials, 3rd edn. McGraw-Hill, Boston
6. Bejan A (1995) Convection heat transfer, 3rd edn. Wiley, New York
7. Hughes TJR (1987) The finite element method: linear static and dynamic finite element analysis. Prentice-Hall Inc, New Jersey
8. Incropera FP, DeWitt DP, Bergman TL, Lavine AS (2007) Introduction to heat transfer, 5th edn. Wiley, Hoboken
9. Jin J (2002) The finite element method in electromagnetics, 2nd edn. Wiley-IEEE Press, New York
10. Kaviany M (2001) Principles of heat transfer. Wiley, New York
11. Lau JH (1993) Thermal stress and strain in microelectronics packaging. Van Nostrand Reinhold, New York
12. Logan DL (2002) A first course in the finite element method, 3rd edn. Brooks/Cole, Pacific Grove
13. Nédélec JC (1980) Mixed finite-elements in $\mathbb{R}^3$. Numerische Mathematik 35:315–341. doi:10.1007/BF01396415
14. Reddy JN, Gartling DK (2000) The finite element method in heat transfer and fluid dynamics, 2nd edn. CRC Press, Boca Raton
15. Roark RJ, Young WC (1982) Formulas for stress and strain, 5th edn. McGraw-Hill, New York
16. Salon SJ (1995) Finite element analysis of electrical machines. Kluwer Academic Publishers, Boston
17. Silvester PP, Ferrari RL (1996) Finite elements for electrical engineers, 3rd edn. Cambridge University Press, New York
18. Yee K (1996) Numerical solution of initial boundary value problems involving Maxwell's equations in isotropic media. IEEE Trans Antennas Propag 14:302–307. doi:10.1109/TAP.1966.1138693
19. Zienkiewicz OC, Taylor RL, Zhu JZ (2005) The finite element method: its basis and fundamentals, 6th edn. Elsevier Butterworth-Heinemann, Burlington

Chapter 4
Optimization Methods for Electromechanical Systems

Optimization methods for electromechanical systems are covered in this chapter with a special emphasis on structural optimization for multiphysics systems. Topology optimization as a material distribution problem is explained in the first section through two basic single physics numerical examples. These initial examples serve to illustrate the optimization process for relatively simple physical systems and provide an opportunity to discuss the computational challenges associated with the method. As a variation on this isotropic material distribution approach to structural design, level set techniques for the topological design of structures are then introduced due to their recent growth as a related research field.

The second section of this chapter is then focused on more traditional parametric size analysis/optimization techniques. These methods are well established and can play an important role in the synthesis of design concepts attained through structural topology optimization. A brief explanation of parametric size analysis is provided including a schematic of a representative structural size optimization example for illustrative purposes.

4.1 Topology Optimization

The field of structural optimization can trace its roots back to Anthony Michell's influential 1904 paper [29] that was focused on determining which frame structures were the most economical in withstanding a balanced system of forces. However, despite Michell's early work, the pace of research in the field of structural optimization did not gain speed until the 1960s when the use of electronic computers became more common. As reviewed in [45], research related to mathematical programming based on sensitivity analysis for structural optimization started to emerge during this time period. By the late 1980s and early 1990s, the field of structural optimization had evolved from a limited focus on shape and size optimization to include topology optimization, as evidenced by early pioneering work [3, 45]. Since then, many

E. M. Dede et al., *Multiphysics Simulation,* Simulation Foundations, Methods and Applications, DOI: 10.1007/978-1-4471-5640-6_4,

applications and numerical strategies for structural topology optimization have been developed, as extensively reviewed in [5, 10, 35, 41].

As mentioned in Chap. 1 of this book, the field of structural topology optimization has grown exponentially in the last two decades to encompass a wide variety of technical topics. Relevant target applications for this technology include structural dynamics [18], compliant mechanisms [30, 48], composite materials [17, 36, 37], sensors and actuators [21, 23, 24, 38, 39], fluid flow structures [2, 6, 31], microfluidic mixers [1], phononic materials [40], and fluid-structure interactions [27, 49]. More recently, research studies related to fully coupled multiphysics systems have also emerged including complex problems involving conjugate heat transfer [12, 14, 26, 50], piezoelectric and fluid-structure interactions [28], and even as many as four physical processes [51].

While the reader is referred to the literature for numerous examples, the goal of this section is to review the general form of the topology optimization problems considered throughout this book. The majority of the structural topology optimization examples presented in this text utilize the solid isotropic material with penalization (SIMP) approach [4, 5] for interpolating the design density variables, although structural optimization using a level set function approach [46] is also briefly reviewed. Prior to the discussion of the level set method, two separate single physics optimization examples for heat conduction and fluid flow are provided in this section as a precursor to the more complex multiphysics studies introduced in Chap. 5. Interestingly, these two single physics systems are used as the foundation for several of the multiphysics investigations that follow.

Optimization Problem Description and Computational Process Overview

In the structural topology optimization design examples presented throughout this book a non-linear optimization problem of the standard form described in [43] is considered,

$$\begin{array}{rlll} \text{Find} & \gamma & & \\ \text{Minimize} & \mathsf{F}_o(\gamma) & & \\ \text{Subject to} & g_i(\gamma) \le 0, & \text{for} & i = 1, \ldots, m \\ & \gamma_{j\_\min} \le \gamma_j \le \gamma_{j\_\max}, & \text{for} & j = 1, \ldots, n, \end{array}$$

where $\mathsf{F}_o(\gamma)$ is a single or multi-term objective function, $g_i(\gamma)$ are behavior constraints, and γ is a vector of scalar design variables (typically associated with the material physical parameters of interest) having lower and upper bounds of $\gamma_{j\_\min}$ and $\gamma_{j\_\max}$, respectively.[1]

[1] Reprinted from Ref. [15], Copyright (2012), with permission from Inderscience Enterprises Limited.

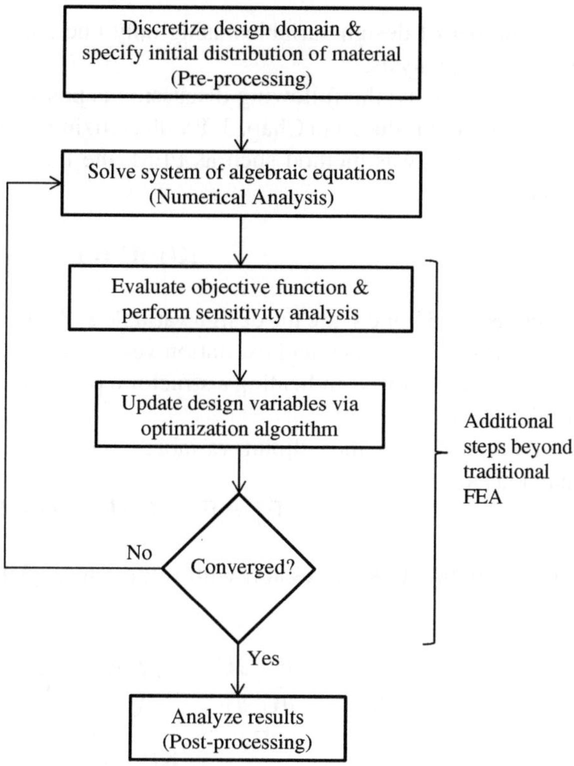

Fig. 4.1 Typical flowchart of computations for structural topology optimization. The additional steps beyond traditional FEA, as described in Chap. 3, are highlighted

A flowchart of computations provided in Fig. 4.1, where the objective function is minimized using its sensitivity with respect to the design variable, γ. In many of the design examples throughout this book, an iterative computational approach that couples in-house or commercial finite element software [9] with an external optimizer is employed in the solution of the optimization problem. The method of moving asymptotes (MMA) optimizer algorithm developed by Svanberg [43] is used in many of the numerical studies herein since it has historically performed well in many topology optimization problems; however, other approaches are possible. During the optimization process, the objective function is repeatedly evaluated by computing the response of a structure using finite element solvers. The sensitivity analysis is then performed at each iteration using either a manually derived sensitivity function or sensitivities computed using the adjoint variable method, per [5]. From there, the design variables are updated via the optimizer. Once a specific convergence criterion is satisfied, the final results are analyzed and the optimal solution may be post-processed.

Adjoint Sensitivity Analysis

The aforementioned adjoint variable method (AVM) is commonly used for sensitivity analysis in topology optimization. The main advantage of AVM is its efficiency. With AVM, it is possible to estimate the sensitivity of the objective function relative to any number of design variables using only one additional numerical analysis, i.e., the adjoint analysis.

For simplicity, the following discussion is posed in terms of the static structural system first introduced in Chap. 3. By discretizing the governing equation with some numerical analysis method such as FEM, the following linear algebra equation is obtained

$$\mathbf{K}(\gamma)\mathbf{U} = \mathbf{F}, \tag{4.1}$$

where $\mathbf{K}(\gamma)$, $\mathbf{U}$, and $\mathbf{F}$ are the design variable embedded system matrix, state variable vector (or field vector), and excitation vector, respectively. Note that, for compactness, the subscript, s, indicating a structural problem in Eq. (3.11) has been dropped in Eq. (4.1).

By introducing the adjoint variable vector, $\hat{\lambda}$, the objective function is augmented as

$$\mathsf{F}_o^* = \mathsf{F}_o - \hat{\lambda}^{\mathrm{T}}(\mathbf{K}(\gamma)\mathbf{U} - \mathbf{F}). \tag{4.2}$$

Differentiating this expression with respect to γ yields the sensitivity of objective function, F_o, as follows

$$\begin{aligned} \frac{\partial \mathsf{F}_o{}^*}{\partial \gamma} &= \frac{\partial \mathsf{F}_o}{\partial \mathbf{U}}\frac{\partial \mathbf{U}}{\partial \gamma} - \hat{\lambda}^{\mathrm{T}}\left(\frac{\partial \mathbf{K}(\gamma)}{\partial \gamma}\mathbf{U} + \mathbf{K}(\gamma)\frac{\partial \mathbf{U}}{\partial \gamma}\right) \\ &= \left(\frac{\partial \mathsf{F}_o}{\partial \mathbf{U}} - \hat{\lambda}^{\mathrm{T}}\mathbf{K}(\gamma)\right)\frac{\partial \mathbf{U}}{\partial \gamma} - \hat{\lambda}^{\mathrm{T}}\left(\frac{\partial \mathbf{K}(\gamma)}{\partial \gamma}\mathbf{U}\right). \end{aligned} \tag{4.3}$$

By choosing $\hat{\lambda}$ to satisfy the following equation, Eq. (4.4), the term $\frac{\partial \mathbf{U}}{\partial \gamma}$ can be eliminated from Eq. (4.3)

$$\frac{\partial \mathsf{F}_o}{\partial \mathbf{U}} - \hat{\lambda}^{\mathrm{T}}\mathbf{K}(\gamma) = 0. \tag{4.4}$$

Once the values of the state variables, $\mathbf{U}$, and adjoint variables, $\hat{\lambda}$, are obtained, they may be substituted into the expression for the design sensitivity,

$$\frac{\partial \mathsf{F}_o{}^*}{\partial \gamma} = -\hat{\lambda}^{\mathrm{T}}\left(\frac{\partial \mathbf{K}(\gamma)}{\partial \gamma}\mathbf{U}\right), \tag{4.5}$$

for any γ using only simple arithmetic operations.

Here, Eq. (4.4) is called the adjoint equation, which characterizes the adjoint problem. Its computational cost is equivalent to normal finite element analysis.

Single Physics Topology Optimization—Heat Conduction

Since many of the multiphysics systems introduced in Chap. 2 involve heat transfer, it is appropriate that the first single physics design optimization example presented below is focused on pure heat conduction, where the governing equation is

$$-\nabla \cdot \left[k(\gamma)\nabla T\right] = Q. \tag{4.6}$$

In Eq. (4.6), T is the temperature state variable, Q is the internal heat generation, and $k(\gamma)$ is the material thermal conductivity, which is a function of the design density variable, γ. The thermal conductivity is then interpolated using a SIMP approach

$$k(\gamma) = (0.001 + 0.999\gamma^{p})k_o \tag{4.7}$$

where k_o is the thermal conductivity of the assumed solid thermally conductive material, and p is a standard penalization parameter set equal to 3 following [5]. In Eq. (4.7), $k(\gamma) \to 0.001k_o$ as $\gamma \to 0$, and $k(\gamma) \to k_o$ as $\gamma \to 1$.

To determine the optimal thermal conduction path, a general energy formulation of the objective function is integrated over the design domain, Ω_d, as follows

$$\mathsf{F}_o = \int_{\Omega_d} \left[\frac{1}{2}k(\gamma)\left(\frac{\partial^2 T}{\partial x^2} + \frac{\partial^2 T}{\partial y^2}\right)\right] \mathrm{d}\Omega_d, \tag{4.8}$$

and the full optimization problem may be formulated as

Find	γ
Minimize	Eq. (4.8)
Subject to	Eq. (4.6)
	$\int_{\Omega_d} \gamma \mathrm{d}\Omega_d - v_u \leq 0$
	$0 \leq \gamma \leq 1$
Given	Eq. (4.7),

where v_u is an upper-bound limit on the allowable solid (i.e., $\gamma \to 1$) material volume.

As mentioned above, in order to compute the optimal topology, the sensitivity of the objective function with respect to the design variable is required. In a typical single physics example, the manual derivation of the sensitivity function, $\partial \mathsf{F}_o/\partial\gamma$, may be straightforward. However, a more general approach to determining the sensitivity of the objective function with respect to the design variable for complex problems is to use the adjoint variable method, as just explained, to numerically compute the sensitivities. A further advantage of the adjoint method is that it can be readily implemented in commercial finite element software programs such as COMSOL®[2] [9], and exhaustive work has been performed to show the capability of this approach for more complicated physical situations [31]. The reader is referred to the literature for greater details and example code in a high-level programming language.

[2] COMSOL® is a registered trademark of COMSOL AB.

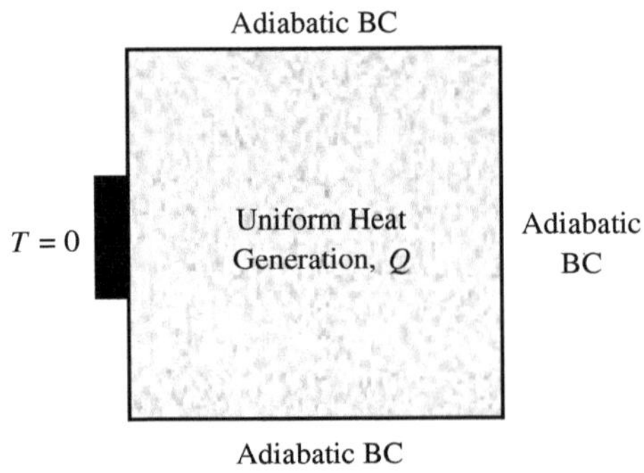

Fig. 4.2 2-D design domain and boundary conditions for a single physics pure heat conduction topology optimization problem

Thus, the optimization process illustrated in Fig. 4.1 was carried out for this example problem using this commercial software [9], and AVM was adopted in computing the sensitivities directly in the program. The MMA algorithm was additionally coupled to the code via a custom MATLAB®[3] [25] script to realize the complete flow of computations (including additional steps beyond traditional FEA), as described in Fig. 4.1.

The assumed 2-D single physics heat conduction model geometry with boundary conditions is shown in Fig. 4.2, where the design domain is 1 m square. A fixed (zero temperature) condition was enforced at the middle of the left edge to represent a heat sink. The remaining edges of the domain were considered adiabatic (i.e., zero heat transfer). Constant uniform heat generation (in the form of a 1 W/m^2 heat flux in 2-D) was applied uniformly to the domain. This model was utilized since it is a standard benchmark in the field of structural topology optimization.

The design domain shown in Fig. 4.2 was meshed using approximately 10,000 quadrilateral elements with 41,000 degrees-of-freedom (DOF). A 40 % solid material volume constraint ($v_u = 0.4$ m^2) was used to obtain the optimal topology shown on the left in Fig. 4.3. Observe that a branching type of topology is constructed which minimizes the conductive thermal resistance between the heat source and sink.

The single physics heat conduction problem is easily extended to 3-D, and a related optimization study was undertaken in [11]. The model geometry and optimal structural topology results for a 40 % solid material volume constraint are shown in Fig. 4.4, for completeness, where yellow colored regions indicate solid material. These results are consistent with those for pure heat conduction found in the literature [7]. However, note that a relatively coarse mesh with 8,000 hexahedral elements and approximately 69,000 DOF was used to generate the structure shown in Fig. 4.4 to avoid incurring a significant penalty in terms of increased computational cost. Thus, the expansion of the optimization approach into the third dimension can quickly lead to massively large-scale systems of equations that must be solved, and this numerical problem is further exacerbated by the inclusion of additional physics.

At this point, it is appropriate to briefly discuss the relationship between mesh size and the structural designs that are obtained via the topology optimization process

[3] MATLAB® is a registered trademark of The MathWorks, Inc.

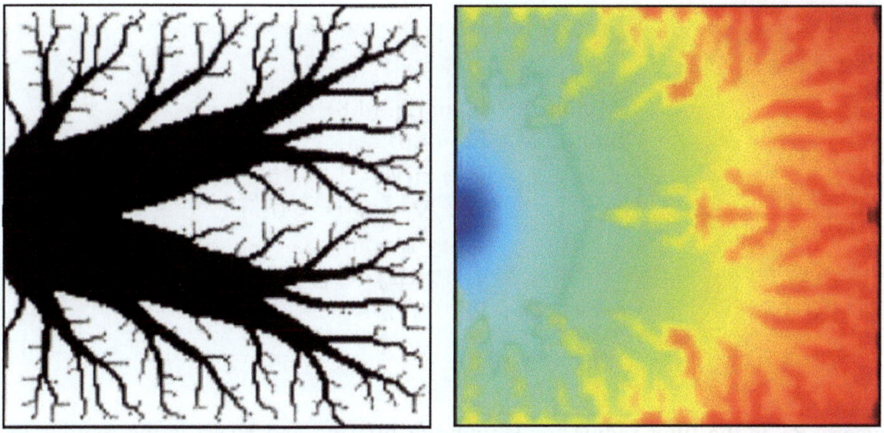

Fig. 4.3 Topology optimization results for the single physics pure heat conduction problem. The optimal topology shown on the left was obtained using a 40 % solid material volume constraint. *Note dark* regions = *solid* (i.e., thermally conductive) material; *light* regions = void (i.e., thermally insulative) material. The image on the *right* shows the computed normalized temperature distribution, where *red* regions indicate a higher temperature

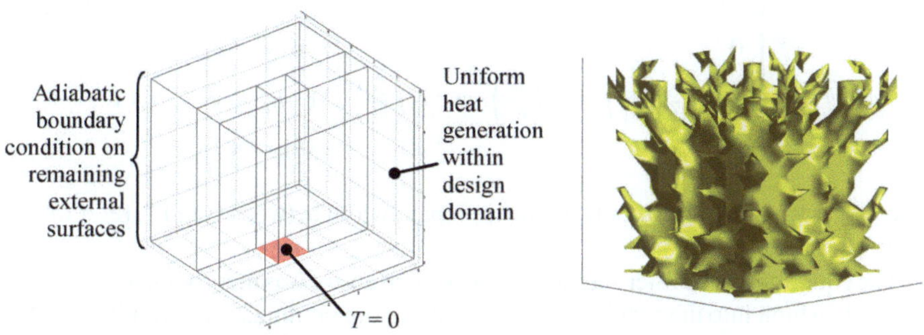

Fig. 4.4 3-D model geometry (*left*) and topology optimization results (*right*) for the single physics pure heat conduction problem [11]. The optimal topology was obtained using a 40 % solid material volume constraint

described thus far. The issue of mesh dependency is explored extensively in the literature [5], and the reader is referred to it for a deep exploration of the topic. However, the simplicity of the 2-D pure heat conduction problem allows for efficient investigation of the basic concept. Generally, finer meshes produce structural designs that exhibit finer features, and this trend continues without limit in the absence of some form of filtering.

Here, a useful Helmholtz PDE filtering routine is adopted [20, 22] to illustrate the manner in which mesh independency and a minimum structural length scale can be efficiently enforced. Specifically, for pure heat conduction, the optimization problem is re-formulated as follows

Find	γ
Minimize	F_o in Eq. (4.8)
Subject to	Eq. (4.6)
	$\gamma = \tilde{\mathsf{H}}(\tilde{\psi})$
	$-R_f^2 \nabla^2 \tilde{\psi} + \tilde{\psi} = \psi$
	$-1 \leq \psi \leq 1$
	$\int_{\Omega_d} \gamma \mathrm{d}\Omega_d - v_u \leq 0$
Given	Eq. (4.7),

where R_f is the filter radius for Helmholtz filtering and $\tilde{\mathsf{H}}(\tilde{\psi})$ is the regularized Heaviside function of the filtered scalar function, $\tilde{\psi}$, with a transition bandwidth of h_t. Note that the design variable has now become ψ, and the original design variable, γ, is now a function of the intermediate projected field, $\tilde{\psi}$. Additionally, the regularized Heaviside function is defined as follows

$$\tilde{\mathsf{H}}(\tilde{\psi}) = \begin{cases} 0 & (\tilde{\psi} < -h_t) \\ \frac{1}{2} + \frac{15}{16}\left(\frac{\tilde{\psi}}{h_t}\right) - \frac{5}{8}\left(\frac{\tilde{\psi}}{h_t}\right)^3 + \frac{3}{16}\left(\frac{\tilde{\psi}}{h_t}\right)^5 & (-h_t \leq \tilde{\psi} \leq h_t)\,, \\ 1 & (h_t < \tilde{\psi}) \end{cases} \tag{4.9}$$

where h_t is a parameter to control transition bandwidth between 0 and 1; refer to [20] for further details.

The equation-based filtering routine introduced above was incorporated in straightforward fashion into the previously described custom optimization script [9, 25]. For this study, the same mesh size as before (i.e., ~10,000 elements) was used for the 2-D heat conduction problem, while the filter radius, R_f, is increased from 2×10^{-5} to 5×10^{-3} and then 2×10^{-2}. The optimal structural topologies obtained using this filtering routine are shown in Fig. 4.5 with increasing R_f values from left to right.

The numerical example serves to illustrate the manner in which a minimum length scale may be enforced even for high levels of mesh refinement. In many cases, such filtering techniques enable higher accuracy solutions while maintaining meaningful structural designs that exhibit greater manufacturability. Achieving minimum length scale, design robustness, and the manufacturing consequences of designs produced using topology optimization are gaining greater momentum as research topics [19, 33], and interesting ideas regarding the implementation of filtering toolboxes for the optimization engineer have been recently proposed [44].

Single Physics Topology Optimization—Fluid Flow

The second single physics design optimization example is focused on the optimization of systems involving Navier–Stokes fluid flow and is motivated by the fact that

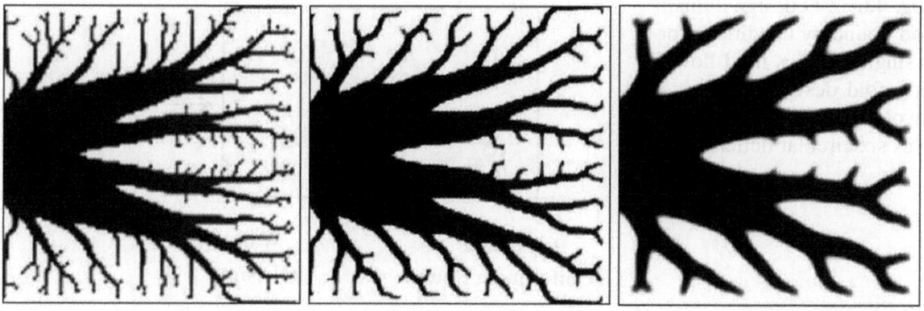

Fig. 4.5 Topology optimization results for the single physics pure heat conduction problem obtained using ~10,000 finite elements, a solid volume of 40 %, and a Helmholtz PDE filtering routine. The results for a filter radius value, R_f, of 2×10^{-5}, 5×10^{-3}, and 2×10^{-2} with a Heaviside function bandwidth of $h_f = 1$ are shown, respectively, at *left*, *center*, and *right*

convective heat transfer frequently plays an important role in the design of many electromechanical devices and components. A method for minimizing the dissipated power in a fluid [6] passing through a flow structure or fluid manifold is adopted. The implementation of the method is based on [31], where fluid flow in a porous medium is assumed, and the final design of a manifold is determined by interpolating the inverse permeability between semi-permeable solid and fluid states. Note that laminar (i.e., deterministic) flow fields with Reynolds (Re) number values below 2100 are assumed in adoption of the method presented below.

The governing equations for steady-state Navier–Stokes laminar fluid flow [34] in the absence of body forces are

$$\nabla \cdot \mathbf{v} = 0, \quad \text{and} \tag{4.10}$$

$$\rho\,(\mathbf{v} \cdot \nabla \mathbf{v}) = -\nabla P + \nabla \cdot \left\{\eta \left[\nabla \mathbf{v} + (\nabla \mathbf{v})^{\mathrm{T}}\right]\right\} - \tilde{\alpha}(\gamma)\mathbf{v}, \tag{4.11}$$

where the fluid density and dynamic viscosity are, respectively, ρ and η, and the inverse permeability of the porous medium is $\tilde{\alpha}(\gamma)$, which is assumed to be approximately valid for an actual porous medium [31]. The state variables in Eqs. (4.10) and (4.11) include the fluid pressure, P, and velocity vector, $\mathbf{v}$.

The effective inverse permeability, $\tilde{\alpha}(\gamma)$, in Eq. (4.11) is a function that is interpolated to define either the solid or fluid regions in the design domain, see [31], where

$$\tilde{\alpha}(\gamma) = \tilde{\alpha}_{\min} + (\tilde{\alpha}_{\max} - \tilde{\alpha}_{\min})\,\frac{q(1-\gamma)}{q+\gamma}. \tag{4.12}$$

In this convex interpolation scheme, q is a tuning parameter set equal to 0.1, while $\tilde{\alpha}_{\min}$ and $\tilde{\alpha}_{\max}$ are the minimum and maximum values of the inverse permeability, respectively.

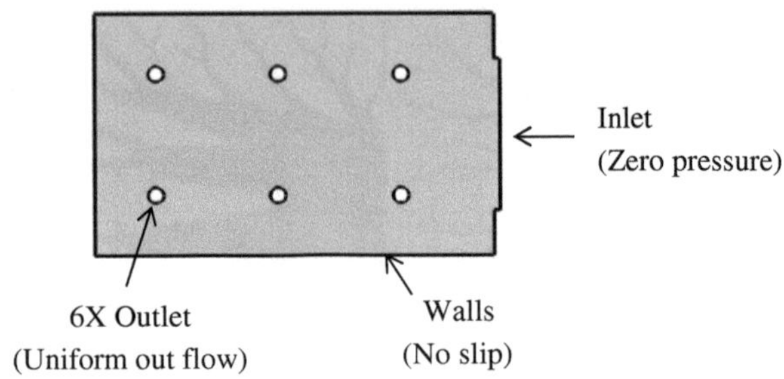

Fig. 4.6 2-D design domain and boundary conditions for a single physics fluid flow manifold design problem involving a single side inlet plus six circular outlets

Following [31], a useful objective function for minimizing the fluid power dissipated (i.e., flow resistance) in an assigned design domain is

$$\mathsf{F}_o = \int_{\Omega_d} \left[\frac{1}{2}\eta \sum_{i,j} \left(\frac{\partial \mathrm{v}_i}{\partial x_j} + \frac{\partial \mathrm{v}_j}{\partial x_i} \right)^2 + \sum_i \tilde{\alpha}(\gamma)\mathrm{v}_i^2 \right] \mathrm{d}\Omega_d, \tag{4.13}$$

and the full optimization problem may be stated as

$$\begin{array}{rl} \text{Find} & \gamma \\ \text{Minimize} & \text{Eq. (4.13)} \\ \text{Subject to} & \text{Eqs. (4.10 – 4.11)} \\ & \int_{\Omega_d} \gamma \mathrm{d}\Omega_d - v_u \leq 0 \\ & 0 \leq \gamma \leq 1 \\ \text{Given} & \text{Eq. (4.12).} \end{array}$$

A simplified 2-D design domain (of size 65 × 40 mm) was assumed for the example single physics fluid flow design optimization problem, as shown in Fig. 4.6. A zero pressure side fluid inlet boundary condition was set on the right side of the design domain, while six circular fluid outlets were distributed uniformly throughout the domain. The flow rate out of each nozzle was assigned the same value in an effort to achieve a relatively uniform flow distribution as opposed to simply assigning a fluid inlet velocity or flow rate condition. A no-slip condition was enforced on the remaining side wall boundaries. The design domain was meshed using ∼4,400 triangular finite elements for a total of approximately 29,200 DOF including the x and y components of the fluid velocity vector plus the pressure state and design optimization variables. For reference, greater details about a related numerical example may be found in [15].

The optimal fluid manifold topology that balances the fluid flow rate out of all six nozzles (with Re ≈ 10) is shown in Fig. 4.7 along with the normalized fluid velocity

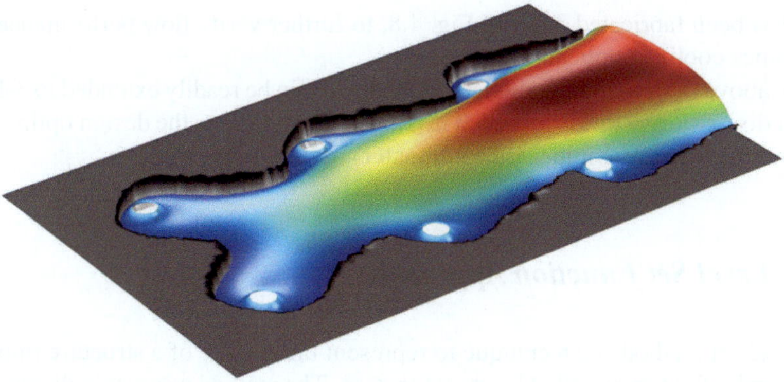

Fig. 4.7 Topology optimization results for the single physics fluid flow manifold design problem. The optimal topology was obtained using a 50 % solid material volume constraint. *Note dark* regions = solid; *light* regions = fluid. The image includes the superimposed fluid velocity contours with larger fluid velocities shown in *red* at higher elevations

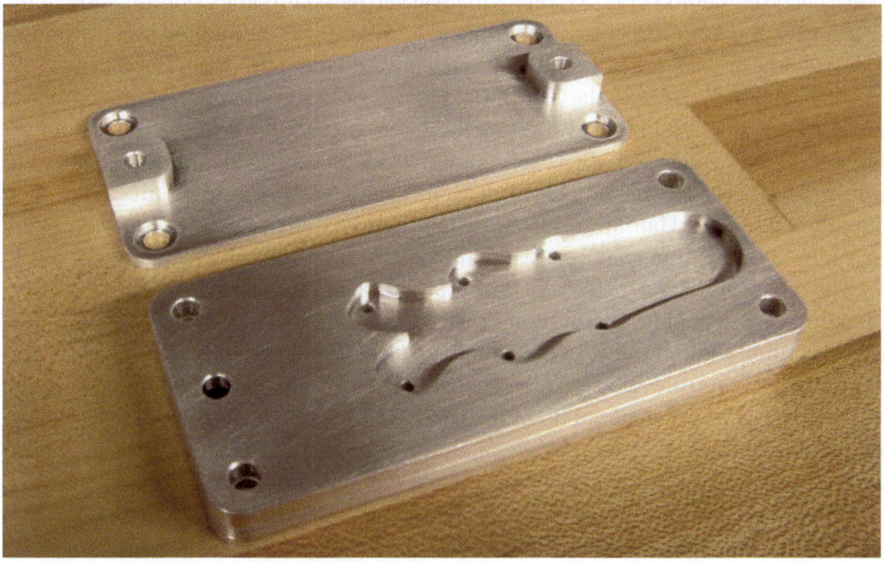

Fig. 4.8 Fabricated manifold prototype based on topology optimization results for the single physics fluid flow design problem. Both the manifold (shown on *bottom*) and cap (shown on *top*) were machined out of aluminum material

contours. Note that larger fluid velocities are shown in red at higher elevations and a 50 % solid material volume constraint was used. The curvy side wall manifold has been compared with traditional manifold designs and was shown to provide reduced flow resistance (pressure drop) and greater outlet flow uniformity, as expected [15]. Additionally, a related prototype structure (designed based on the optimization concept) has been fabricated, refer to Fig. 4.8, to further verify flow performance in an electronics cooling application [13, 16].

The above design optimization method may again be readily extended to 3-D, and further discussion of this is provided in Chap. 5 in relation to the design optimization of thermal-fluid (i.e., conjugate heat transfer) multiphysics systems.

4.1.1 Level Set Function Approach

The level set method is a technique to represent the surface of a structure implicitly with a scalar field or so-called level set function. The method was originally proposed by Osher and Sethian [32], and it has been applied to various problems that involve time dependent varying free surfaces, such as two-phase flow problems. Also, the method may be applied to various static problems and structural optimization is one of them, that is, what we call level set-based topology optimization. Hereafter, the term, level set method or LSM, refers to level set-based topology optimization in this section. While both topology optimization and the level set method were proposed independent of each other, both techniques have a large part in common.

Both methods use a scalar field, the density field in topology optimization and the level set function in the level set method, to represent the designed structure, and the field value is updated after each numerical iteration based on design sensitivity information. Usually, the level set function is updated by solving a time-dependent equation, a kind of Hamilton–Jacobi equation or so-called level set equation, while topology optimization usually uses mathematical programming techniques.

The most significant difference is interpretation of the scalar field. Topology optimization directly interprets the field distribution as the shape and the topology of the structure, while the level set method extracts the iso-contour, usually at a zero value (i.e., the zero-level set) as the surface of the structure thus regarding one side of the contour as the material domain. Theoretically, this means that there exists no intermediate material state in between air and solid (unlike topology optimization), and thus, the level set method is free of the so-called gray scale problem.

Figure 4.9 illustrates the overall idea of the level set method. The level set function, which is shown in the colored height map in the figure, is updated as time progresses. There is a black line on the level set function which indicates the iso-contour at a certain value such as zero. At the bottom, the iso-contour is projected on the plane and inside of the contour is filled to indicate that material domain which is defined by the level set function. As the level set function is updated, the material domain also is updated. Sometimes, this update process involves topological changes, such as splitting or merging of the domains or dissipation of holes.

Algorithm of the Level Set Method

Hereafter, a brief introduction of a level set-based topology optimization method described in a paper by Yamasaki et al. [47] is given.

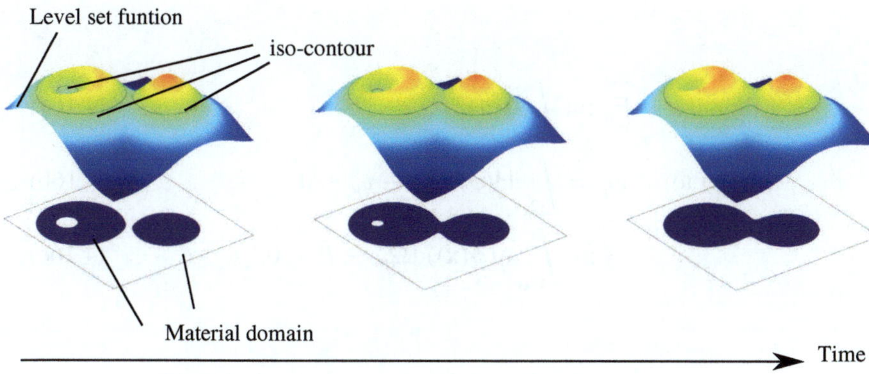

Fig. 4.9 Time-dependent update of level set function and represented shape

Let Ω_d represent the fixed design domain, which entirely envelopes the structure to be designed. The designed structure is filled with a material, so that it is equivalent to the material domain, Ω_m, and the complementary domain is the void domain. The interface of the two domains is defined by the level set contour of the level set function, $\phi(\mathbf{x})$

$$\begin{cases} \phi(\mathbf{x}) > 0 & \text{for} \quad \forall \mathbf{x} \in \Omega_m \setminus \partial\Omega_m, \\ \phi(\mathbf{x}) = 0 & \text{for} \quad \forall \mathbf{x} \in \partial\Omega_m, \\ \phi(\mathbf{x}) < 0 & \text{for} \quad \forall \mathbf{x} \in \Omega_d \setminus \Omega_m, \end{cases} \tag{4.14}$$

where $\mathbf{x}$ stands for a position in Ω_d, and $\partial\Omega_m$ is the boundary of Ω_m. By using the above level set function, an arbitrary shape and topology of the material domain, Ω_m, in the fixed design domain, Ω_d, can be represented.

The level set function is then projected to material density distribution by the following Heaviside function,

$$\mathsf{H}(\phi) = \begin{cases} 0 & (\phi < 0) \\ 1 & (\phi \geq 0). \end{cases} \tag{4.15}$$

Since the above equation is non-differentiable, it is not appropriate in present form for numerical treatment, especially sensitivity analysis, where the differentiation of the objective function with respect to the level set function is performed. As mentioned in [47], several workarounds are proposed for this, where we may use a regularized Heaviside function, $\tilde{\mathsf{H}}(\phi)$; refer to the previous section for the definition of this function.

Here, let us assume that the following structural optimization problem is solved with the level set method

$$\underset{\phi}{\text{Minimize}} \quad \mathsf{F}_o := \int_{\Omega_d} l(u) \mathrm{d}\Omega_d \tag{4.16a}$$

$$\text{Subject to} \quad g_1 := \int_{\Omega_d} \tilde{\mathsf{H}}(\phi) \mathrm{d}\Omega_d - v_u \leq 0 \tag{4.16b}$$

$$g_2 := \int_{\Omega_m} \delta(\phi(\mathbf{x})) \mathrm{d}\Omega_m - \bar{P} < 0, \tag{4.16c}$$

where l is the compliance of the structure, u is the physical state variable, v_u is upper bound of material volume, and $\bar{P}$ is upper bound for the perimeter of the structure for regularization of the complexity of the shape (i.e., so-called perimeter control).

With the classic formulation, the level set function is updated with the following level set equation

$$\frac{\partial \phi\,(\mathbf{x}, t)}{\partial t} + \mathrm{v}_N(\mathbf{x}, t)\,|\nabla\phi(\mathbf{x}, t)| = 0, \tag{4.17}$$

where t is fictitious time and $\mathrm{v}_N(\mathbf{x}, t)$ is the normal velocity provided by sensitivity analysis. Solving Eq. (4.17) is not a trivial task since it may contain singular points. Therefore, Yamasaki et al. [46] proposed a technique called geometrical re-initialization, which extracts the level set contour and calculates the signed distance from each node to the level set contour geometrically. With this method, $|\nabla\phi(\mathbf{x}, t)| = 1$ is always maintained, and thus, Eq. (4.17) is reduced to a simpler form

$$\frac{\partial \phi\,(\mathbf{x}, t)}{\partial t} + \mathrm{v}_N(\mathbf{x}, t) = 0. \tag{4.18}$$

Then, this expression is discretized with respect to t using forward finite difference

$$\frac{\phi(\mathbf{x}, t + \Delta t) - \phi(\mathbf{x}, t)}{\Delta t} + \mathrm{v}_N(\mathbf{x}, t) = 0, \tag{4.19}$$

and finally, this yields the following recursive update equation

$$\phi(\mathbf{x}, t + \Delta t) = \phi(\mathbf{x}, t) - \Delta t \mathrm{v}_N(\mathbf{x}, t). \tag{4.20}$$

Since the level set method itself does not have a framework to handle constraints, the Lagrangian is built with the objective function and constraints

$$\mathsf{L}(\phi) = \mathsf{F}_o(\phi) + \sum_i \hat{\lambda}_i g_i(\phi), \tag{4.21}$$

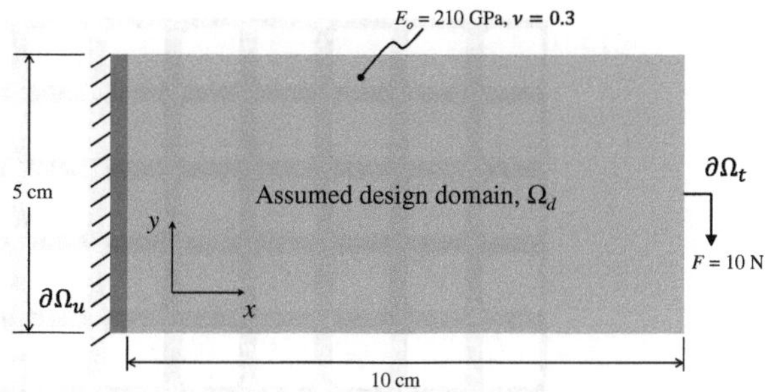

Fig. 4.10 Boundary conditions for a simple plane-stress cantilevered beam numerical example

where $\hat{\lambda}_i$ is the Lagrange multiplier for the constraint, g_i. A natural extension of this method is to use the augmented Lagrangian method for update of the Lagrange multipliers.

The normal velocity is calculated using design sensitivity information of the Lagrangian

$$v_N = \frac{\partial \mathrm{L}(\phi)}{\partial \phi} \tag{4.22}$$

and the update time step is limited by the Courant–Friedrichs–Lewy condition

$$\Delta t = \frac{\alpha_{ls}\Delta x}{||v_N||_{\infty}}, \quad 0 < \alpha_{ls} \leq 1, \tag{4.23}$$

where Δx is the maximum distance of adjacent mesh nodes; refer to [47] for additional details.

Level Set Method – Structural Example

Consider a plane-stress short cantilever benchmark problem to minimize compliance to the boundary condition depicted in Fig. 4.10. The BC, $\partial\Omega_u$, is fixed and a 10 N load is vertically applied downward on $\partial\Omega_t$. The elastic modulus of the solid portion of the material domain is $E_o = 2.1 \times 10^{11}$ N/m^2 and the Poisson's ratio is $\nu = 0.3$. The allowed volume fraction of the material domain is 0.5. The level set method cannot be started with a completely flat structure. So, a trivial initial design is given, as shown in Figs. 4.11 and 4.12 shows the optimal configuration obtained with the level set method [47]. As shown in the figure, the method yields a reasonable black and white result.

4.2 Parametric Size Analysis

In the above section of this chapter, an emphasis was placed on the topology optimization of distributed parameter (or continuum) systems, which is made mathematically feasible through the breakdown of the continuum into many discrete finite elements

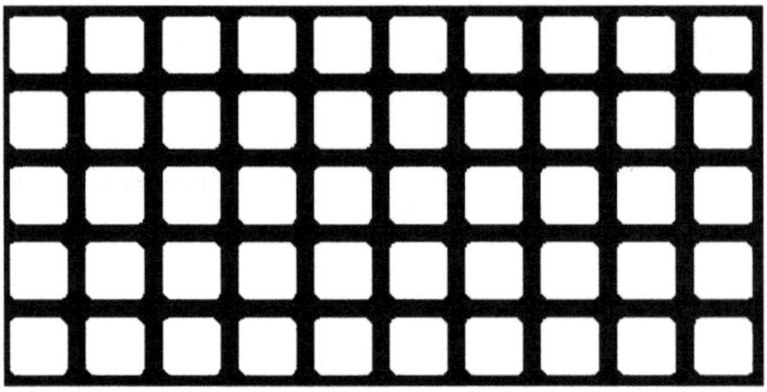

Fig. 4.11 Provided structure (for the level set method algorithm). With kind permission from Springer Science+Business Media: Structural and Multidisciplinary Optimization, A level set-based topology optimization method using the discretized signed distance function as the design variables, Yamasaki et al. [47, Fig. 4], © Springer-Verlag 2009

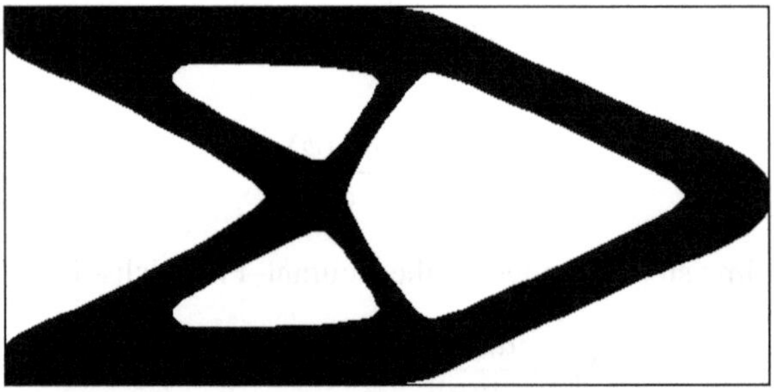

Fig. 4.12 Optimal structure of the example for the 2-D minimum compliance problem. Perimeter constraint is imposed, initial structure is Fig. 4.11 and number of elements = 100 × 50. With kind permission from Springer Science+Business Media: Structural and Multidisciplinary Optimization, A level set-based topology optimization method using the discretized signed distance function as the design variables, Yamasaki et al. [47, Fig. 7b], © Springer-Verlag 2009

using either home-grown or commercial FEA code coupled with a gradient-based optimizer. In contrast, parametric size optimization is related to systems that may be characterized initially by way of a set of discrete parameters prior to further discretization that may be required for analysis purposes. Classic examples of the optimization of discrete parameter structures such as two- or three-bar trusses subject to a stress, displacement, or stiffness constraint may be found in [8]; refer to Fig. 4.13 for an example commonly found in the literature [43], where a segmented cantilevered beam structure is subject to a displacement constraint. For such problems involving a small set of design parameters (e.g., the two side lengths, x_s, and thickness, t_s, of each section shown in Fig. 4.13), analytical solutions are readily derived. However, as the number of discrete parameters is increased, the need for an efficient numerical approach becomes more significant.

The history and details behind a variety of mathematical programming techniques useful for large-scale discrete parameter systems is covered extensively in the literature [8, 42], and readers are referred to such texts for in-depth coverage. Here, the

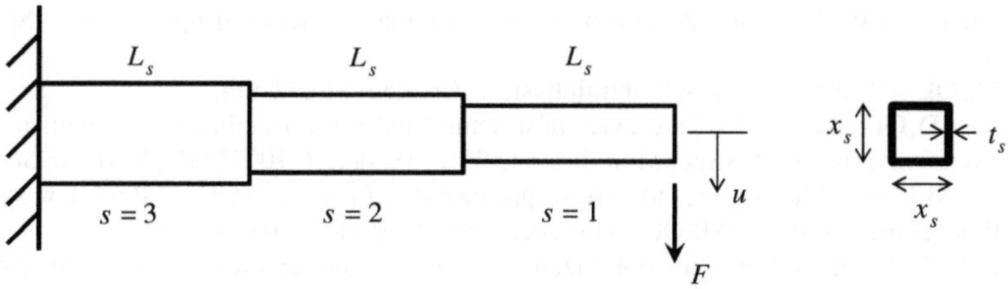

Fig. 4.13 A cantilevered beam structure consisting of three segments each having a hollow square cross section. The beam consists of m segments each of length L_s. Each segment has a thickness, t_s, with a side length of x_s for segment $s = 1, \ldots, m$ for the design variables, as shown in the beam end view image on the *right*

main focus in this book is on the application of automated parametric sweeps to the informed design and structural optimization of 3-D multiphysics systems involving multiple discrete variables (related to the physical size of components) within a finite element framework. While such tools do not necessarily employ gradient-based optimizers to arrive at a globally optimal structural configuration, they allow for rapid visualization and performance comparison of a range of possible designs that are selectable based on one or more other determining factors such as global manufacturability, cost, or packaging constraints. These tools are shown to complement the structural topology optimization techniques described earlier in this chapter. Two representative multiphysics parametric sizing analysis/optimization studies are presented in Chap. 5 in relation to the composite laminate layer thickness determination of a thermal-structural substrate assembly for an electronics packaging application and the selection of a specific channel aspect ratio for an electronics cold plate assembly.

References

1. Andreasen CS, Gersborg AR, Sigmund O (2008) Topology optimization for microfluidic mixers. Int J Numer Meth Fl 61:498–513. doi:10.1002/fld.1964
2. Aage N, Poulsen TH, Gersborg-Hansen A, Sigmund O (2008) Topology optimization of large scale Stokes flow problems. Struct Multidiscip O 35:175–180. doi:10.1007/s00158-007-0128-0
3. Bendsøe MP, Kikuchi N (1988) Generating optimal topologies in structural design using a homogenization method. Comput Method Appl M 71:197–224. doi:10.1016/0045-7825(88)90086-2
4. Bendsøe MP, Sigmund O (1999) Material interpolation schemes in topology optimization. Arch Appl Mech 69:635–654. doi:10.1007/s004190050248
5. Bendsøe MP, Sigmund O (2003) Topology optimization: theory, methods, and applications, 2nd edn. Springer, Berlin
6. Borvall T, Petersson J (2003) Topology optimization of fluids in Stokes flow. Int J Numer Meth Fl 41:77–107. doi:10.1002/fld.426
7. Chen Y, Zhou S, Li Q (2010) Multiobjective topology optimization for finite periodic structures. Comput Struct 88:806–811. doi:10.1016/j.compstruc.2009.10.003

8. Christensen PW, Klarbring A (2009) An introduction to structural optimization. Springer, Berlin
9. COMSOL AB (2008) COMSOL multiphysics, Ver. 3.5a. Stockholm.
10. Deaton JD, Grandhi RV (2013) A survey of structural and multidisciplinary continuum topology optimization: post 2000. Struct Multidiscip O 49:1–38. doi:10.1007/s00158-013-0956-z
11. Dede EM (2009) Multiphysics topology optimization of heat transfer and fluid flow systems. In: Proceedings of the COMSOL conference 2009, Boston, 8–10 Oct 2009
12. Dede EM (2010) Multiphysics optimization, synthesis, and application of jet impingement target surfaces. In: Proceedings of the 12th IEEE intersociety conference on thermal and thermomechanical phenomena in electronic systems, Las Vegas, 2–5 June 2010. doi:10.1109/ITHERM.2010.5501408
13. Dede EM, Liu Y (2011) Heat exchanger fluid distribution manifolds and power electronics modules incorporating the same. US Patent 8,077,460 B1
14. Dede EM (2012) Optimization and design of a multipass branching microchannel heat sink for electronics cooling. J Electron Packag 134:041001. doi:10.1115/1.4007159
15. Dede EM, Lee J, Liu Y, Robert B, Yönak SH (2012) Computational methods for the optimization and design of electromechanical vehicle systems. Int J Vehicle Des 58:159–180. doi:10.1504/IJVD.2012.047383
16. Dede EM, Liu Y (2013) Cold plate assemblies and power electronics modules. US Patent, 8,427,832 B2
17. Dede EM, Nomura T, Lee J (2014) Thermal-composite design optimization for heat flux shielding, focusing, and reversal. Struct Multidiscip O 49:59–68. doi:10.1007/s00158-013-0963-0
18. Díaz AR, Kikuchi N (1992) Solutions to shape and topology eigenvalue optimization problems using a homogenization method. Int J Numer Meth Eng 35:1487–1502. doi:10.1002/nme.1620350707
19. Guest JK, Prévost JH, Belytschko T (2004) Achieving minimum length scale in topology optimization using nodal design variables and projection functions. Int J Numer Meth Eng 61:238–254. doi:10.1002/nme.1064
20. Kawamoto A, Matsumori T, Yamasaki S, Nomura T, Kondoh T, Nishiwaki S (2011) Heaviside projection based topology optimization by a PDE-filtered scalar function. Struct Multidiscip O 44:19–24. doi:10.1007/s00158-010-0562-2
21. Kim JE, Kim DS, Ma PS, Kim YY (2010) Multi-physics interpolation for the topology optimization of piezoelectric systems. Comput Method Appl M 199:3153–3168. doi:10.1016/j.cma.2010.06.021
22. Lazarov BS, Sigmund O (2011) Filters in topology optimization based on Helmholtz-type differential equations. Int J Numer Meth Eng 86:765–781. doi:10.1002/nme.3072
23. Lee J, Kikuchi N (2010) Structural topology optimization of electrical machinery to maximize stiffness with body force distribution. IEEE T Magn 46:3790–3794. doi:10.1109/TMAG.2010.2052365
24. Lee J, Dede EM, Nomura T (2011) Simultaneous design optimization of permanent magnet, coils, and ferromagnetic material in actuators. IEEE T Magn 47:4712–4716. doi:10.1109/TMAG.2011.2160870
25. MATLAB (2009) The MathWorks Inc, Ver. 7.9. Natick
26. Matsumori T, Kondoh T, Kawamoto A, Nomura T (2013) Topology optimization for fluid-thermal interaction problems under constant input power. Struct Multidiscip O 47:571–581. doi:10.1007/s00158-013-0887-8
27. Maute K, Allen M (2004) Conceptual design of aeroelastic structures by topology optimization. Struct Multidiscip O 27:27–42. doi:10.1007/s00158-003-0362-z
28. Maute K (2014) Topology optimization of coupled multi-physics problems. In: Rozvany GIN, Lewiński T (eds) Topology optimization in structural and continuum mechanics. Springer, Vienna, p 421–437. doi:10.1007/978-3-7091-1643-2_18
29. Michell AGM (1904) The limits of economy of material in frame-structures. Philos Mag 47:589–597. doi:10.1080/14786440409463229

30. Nishiwaki S, Frecker MI, Min S, Kikuchi N (1998) Topology optimization of compliant mechanisms using the homogenization method. Int J Numer Meth Eng 42:535–559. doi:10.1002/(SICI)1097-0207(19980615)42:3<535:AID-NME372>3.0.CO;2-J
31. Olesen LH, Okkels F, Bruus H (2006) A high-level programming-language implementation of topology optimization applied to steady-state Navier-Stokes flow. Int J Numer Meth Eng 65:975–1001. doi:10.1002/nme.1468
32. Osher S, Sethian JA (1988) Fronts propagating with curvature-dependent speed: algorithms based on Hamilton-Jacobi formulations. J Comput Phys 79:12–49. doi:10.1016/0021-9991(88)90002-2
33. Qian X, Sigmund O (2013) Topological design of electromechanical actuators with robustness toward over- and under-etching. Comput Method Appl M 253:237–251. doi:10.1016/j.cma.2012.08.020
34. Reddy JN, Gartling DK (2000) The finite element method in heat transfer and fluid dynamics, 2nd edn. CRC Press, Boca Raton
35. Rozvany GIN (2009) A critical review of established methods of structural topology optimization. Struct Multidiscip O 37:217–237. doi:10.1007/s00158-007-0217-0
36. Sigmund O, Torquato S (1996) Composites with extremal thermal expansion coefficients. Appl Phys Lett 69:3203. doi:10.1063/1.117961
37. Sigmund O, Torquato S (1999) Design of smart composite materials using topology optimization. Smart Mater Struct 8. doi:10.1088/0964-1726/8/3/308
38. Sigmund O (2001) Design of multiphysics actuators using topology optimization - Part I: one-material structures. Comput Method Appl M 190:6577–6604. doi:10.1016/S0045-7825(01)00251-1
39. Sigmund O (2001) Design of multiphysics actuators using topology optimization - Part II: two-material structures. Comput Method Appl M 190:6605–6627. doi:10.1016/S0045-7825(01)00252-3
40. Sigmund O, Jensen JS (2003) Systematic design of phononic band-gap materials and structures by topology optimization. Philos T Roy Soc A 361:1001–1019. doi:10.1098/rsta.2003.1177
41. Sigmund O, Maute K (2013) Topology optimization approaches. Struct Multidiscip O 48:1031–1055. doi:10.1007/s00158-013-0978-6
42. Spillers WR, MacBain KM (2009) Structural optimization. Springer, Dordrecht
43. Svanberg K (1987) The method of moving asymptotes—a new method for structural optimization. Int J Numer Meth Eng 24:359–373. doi:10.1002/nme.1620240207
44. Svanberg K, Svärd H (2013) Density filters for topology optimization based on the Pythagorean means. Struct Multidiscip O 48:859–875. doi:10.1007/s00158-013-0938-1
45. Suzuki K, Kikuchi N (1991) A homogenization method for shape and topology optimization. Comput Method Appl M 93:291–318. doi:10.1016/0045-7825(91)90245-2
46. Yamasaki S, Nishiwaki S, Yamada T, Izui K, Yoshimura M (2010) A structural optimization method based on the level set method using a new geometry-based re-initialization scheme. Int J Numer Meth Eng 83:1580–1624. doi:10.1002/nme.2874
47. Yamasaki S, Nomura T, Kawamoto A, Sato K, Izui K, Nishiwaki S (2010) A level set based topology optimization method using the discretized signed distance function as the design variables. Struct Multidiscip O 41:685–698. doi:10.1007/s00158-009-0453-6
48. Yin L, Ananthasuresh GK (2002) A novel topology design scheme for the multi-physics problems of electro-thermally actuated compliant micromechanisms. Sensor Actuat A-Phys 97–98:599–609. doi:10.1016/S0924-4247(01)00853-6
49. Yoon GH, Jensen JS, Sigmund O (2007) Topology optimization of acoustic-structure interaction problems using a mixed finite element formulation. Int J Numer Meth Eng 70:1049–1075. doi:10.1002/nme.1900
50. Yoon GH (2010) Topological design of heat dissipating structure with forced convective heat transfer. J Mech Sci Technol 24:1225–1233. doi:10.1007/s12206-010-0328-1
51. Yoon GH (2012) Topological layout design of electro-fluid-thermal-compliant actuator. Comput Method Appl M 209–212:28–44. doi:10.1016/j.cma.2011.11.005

Chapter 5
Electromechanical System Simulation and Optimization Studies

Several multiphysics simulation and design optimization example studies for electromechanical systems are provided in this chapter. Examples related to electronic system components are covered first followed by magnetic components, RF devices, actuators, and motors. Building off of the background information provided in Chaps. 3 and 4, the governing equations for each problem at hand are reintroduced with the addition of the defined optimization variables.

In Sect. 5.1, five different electronic system component analysis and design studies are introduced in order of increasing complexity. Here, we generally equate greater problem complexity with one of the following features:

1. Multiphysics systems with governing equations that have different forms;
2. Optimization in three dimensions versus two dimensions;
3. An increased number of physical processes, design objectives, or constraints;
4. The addition of material microstructure considerations.

In Sect. 5.1.1, the optimization of electrothermal systems is presented with a focus on the 2-D design of electrical conductors subject to design-dependent cooling for electrical and thermal performance objectives. Next, the design optimization of 3-D thermal-structural systems is covered in Sect. 5.1.2, where the optimal design of planar substrate materials for electronics packaging applications is considered in order to reduce thermally induced stress effects. The design optimization of thermal-fluid systems in both 2-D and 3-D design domains is then presented in Sect. 5.1.3 in the context of cold plate design for high heat flux electronic systems. A derived 2-D cooling channel topology is explored in greater depth through subsequent parametric analysis of the resultant finned heat sink design, and some example prototype structures are shown. The optimization of thermal-fluid systems is then extended to a unique case involving three highly coupled physical processes in Sect. 5.1.4, where the motion of a magnetically susceptible cooling fluid is controlled through the design of a magnetic field source. This latter example illustrates the utility of structural optimization techniques in arriving at nonintuitive concepts for performance enhancement of complex multiphysics systems. Finally, heat flow control in anisotropic materials is included as a supplementary topic in Sect. 5.1.5, where

E. M. Dede et al., *Multiphysics Simulation,* Simulation Foundations, Methods and Applications, DOI: 10.1007/978-1-4471-5640-6_5,

the material microstructure is specified using a composite micromechanics model to achieve unique thermal performance characteristics. In this last example study, some structural embodiments of such materials for electronics applications are provided.

Moving to magnetic components in Sect. 5.2, computational examples for the analysis and design of these components are presented including a representative inductor analysis and a related topology optimization study. From there, a higher frequency RF microstrip device example and a multiphysics antenna design study are explained in Sect. 5.3. In the second RF case, a sequentially solved multiphysics system is optimized with respect to both electromagnetic and structural response. In addition, the multiphysics antenna design problem is expanded to consider the effect of fabrication constraints on the final optimized device.

Concluding this chapter, the design optimization of actuators and motors is, respectively, covered in Sects. 5.4 and 5.5. The magnetostructural design of a basic solenoid actuator is first introduced followed by a more sophisticated linear actuator structural topology optimization example involving several different materials (e.g., the actuator coil, PM, and iron) and hence design variables. The application of multiphysics simulation and optimization tools to the design of more complicated switched reluctance motors is then described in the last section of the chapter along with the performance analysis of a more traditional interior permanent magnet synchronous motor.

5.1 Electronic System Component Analysis and Design

Thermal management of electronic systems is crucial for their reliable operation and efficient design tools for heat sinks, cold plates, bus bars, and packaging materials are increasingly becoming important as assembly size decreases and power density increases. Accordingly, five design optimization examples are covered below with the first topic centered on exploring the trade-offs between minimizing electrical resistance versus thermal resistance in the multiobjective design optimization of electrical conductors. Thermomechanical systems are then addressed in the next section, where a direct-bonded copper (DBC) substrate composite is designed for increased reliability by minimizing the maximum stress present at the bonded interface between the substrate and an attached electronics device during high temperature operation. From there, the focus shifts to traditional conjugate heat transfer thermal-fluid systems and applications to cold plate design, and building on this, a more complex three-physics fully coupled system is subsequently proposed, which involves the design layout of a heat sink for a localized heat source considering thermomagnetic convective heat transfer. In the final part of this section, heat flow control in structures that exhibit anisotropic material physical properties (specifically, thermal conductivity) is introduced to illustrate how structural optimization methods may be applied at multiple physical length scales. From a broad perspective, the different simulation and optimization techniques presented in this section represent a suite of numerical tools for enhanced electro-thermomagnetic system concept development and design.

Logically, many of these tools may be synthesized and/or adapted to other related engineering applications.

5.1.1 Design Optimization of Electrothermal Systems

Electromechanical systems found in today's power conversion applications (e.g., hybrid vehicles, photovoltaic systems, and wind turbines) involve electrical conductor components such as fuses and bus bars that carry high currents [38]. In the case of bus bars, a metal strip is routed in lieu of an electrically conducting cable to carry the large current over short distances. As explained in [38], bus bars with different voltage potentials are often placed in close proximity to each other, and a thin layer of dielectric material is commonly positioned between bus bars to assure electrical isolation. Consequently, thermal management of bus bars is critical since hot spots generated by the Joule heating of the conductor can potentially damage this electrical isolation barrier and cause severe short circuits.

Multiphysics simulation is a readily available tool for better understanding of the Joule heating phenomenon that occurs in the operation of a bus bar, and a multiobjective topology optimization approach may be utilized in conjunction to understand the trade-off between a more efficient electrical conductor versus a more efficient thermal conductor. Accordingly, in this section, the basic design of a conductor considering both the minimization of electrical and thermal resistance is presented, where heat loss to the environment is addressed through the application of a surface convection coefficient at the solid-to-air interface. Since the solid–air boundary of the electrical conductor is dependent on the topological design of the conductor, the surface convection coefficient must track this interface and thus becomes a design-dependent load following the approach outlined in [44]. For simplicity, the topology optimization of a conductor subject to Joule heating is presented in 2-D to illustrate basic trade-offs between the different objectives and the significance of the design-dependent loads.

Building off of the introduction in Sect. 3.2, the governing equations for the Joule heating of a conductor (now with side/surface convection as described in [86]) are given as

$$\rho C_p \frac{\partial T}{\partial t} = \nabla \cdot \left[k(\gamma) \nabla T \right] - h(\gamma)(T - T_o) + Q, \tag{5.1}$$

$$\nabla \cdot \mathbf{J} = Q_j, \tag{5.2}$$

$$\mathbf{J} = \varsigma(\gamma)\mathbf{E} + \mathbf{J}_e, \tag{5.3}$$

$$\mathbf{E} = -\nabla \Phi, \tag{5.4}$$

where the state variables include the conductor temperature, T, current density, $\mathbf{J}$, and electric scalar potential, Φ. In Eqs. (5.1) and (5.3), the conductor-specific heat capacity, thermal conductivity, and electrical conductivity are given as C_p, $k(\gamma)$, and

$\varsigma(\gamma)$, where the latter two material physical parameters are defined as a function of the design density variable, γ. The heat source term, Q, in Eq. (5.1) arises from the electric current determined via Eqs. (5.2)–(5.4), where $\mathbf{J}_e$ and $\mathbf{E}$ are the externally applied current and electric field strength, respectively. For steady-state analysis (as considered below), the term on the left-hand side of Eq. (5.1) disappears.

In the bus bar design optimization problem, both the thermal conductivity and electrical conductivity material physical parameters are interpolated using a SIMP approach to determine the design of the conductor as follows:

$$k(\gamma) = (0.001 + 0.999\gamma^p)k_o, \quad \text{and} \tag{5.5}$$

$$\varsigma(\gamma) = (0.001 + 0.999\gamma^p)\varsigma_o, \tag{5.6}$$

where k_o and ς_o are the thermal conductivity and electrical conductivity of the assumed solid conductor material, and p is a standard penalization parameter set equal to three following [6].

Since bus bars are typically not completely covered by a thermal/electrical insulation layer, convection to the environment may occur through the external surface of the conductor, and this effect plays a role in heat transfer during operation. In the design optimization example that follows, this convective heat transfer effect is captured by enforcing a surface convection heat transfer boundary condition, refer to Sect. 3.2, at the solid-to-void interface of the conductor. Here, an approach similar to [44] is adopted, where the design-dependent heat transfer coefficient, $h(\gamma)$, is defined via a smoothed hat-function as

$$h(\gamma) = \begin{cases} h_o \times 10^{-2} & \text{if } \gamma < \gamma_l - \varepsilon_b \\ h_o\left[\frac{1}{2} + \frac{\gamma-\gamma_l}{2\varepsilon_b} + \frac{1}{2\pi}\sin\left(\frac{\pi(\gamma-\gamma_l)}{\varepsilon_b}\right)\right] & \text{if } \gamma_l - \varepsilon_b < \gamma < \gamma_l + \varepsilon_b \\ h_o & \text{if } \gamma_l + \varepsilon_b < \gamma < \gamma_u - \varepsilon_b \\ h_o\left[\frac{1}{2} - \frac{\gamma-\gamma_u}{2\varepsilon_b} - \frac{1}{2\pi}\sin\left(\frac{\pi(\gamma-\gamma_u)}{\varepsilon_b}\right)\right] & \text{if } \gamma_u - \varepsilon_b < \gamma < \gamma_u + \varepsilon_b \\ 0 & \text{if } \gamma_u + \varepsilon_b < \gamma \end{cases} \tag{5.7}$$

In Eq. (5.7), h_o is a reference full magnitude convective heat transfer coefficient defined at the external surface of the electrical conductor, γ_l and γ_u are, respectively, the lower and upper cutoff values for the surface convection coefficient smoothed hat-function, and ε_b are the size of the bandwidth for the numerical smearing, per [44]. Thus, the surface convection takes a full magnitude value at the solid-to-air interface ($\gamma_l < \gamma < \gamma_u$), while outside of the conductor ($\gamma < \gamma_l$) the surface convection coefficient is a small fraction of this value, and inside the conductor ($\gamma > \gamma_u$) the surface convection coefficient goes to zero. Note that this convective heat transfer effect is implemented as a γ-dependent absorption term, i.e., the second-term on the right-hand side of Eq. (5.1), where T_o is the surface convection reference temperature.

The optimal conductor topology is found by minimizing a multiterm objective function as follows:

$$\mathsf{F}_o = w_1 A + w_2 B, \quad \text{where} \tag{5.8}$$

$$A = \int\limits_{\Omega_d} \left[\frac{1}{2} k(\gamma) \left(\frac{\partial^2 T}{\partial x^2} + \frac{\partial^2 T}{\partial y^2} \right) \right] \mathrm{d}\Omega_d, \quad \text{and} \tag{5.9}$$

$$B = \int\limits_{\Omega_d} \left[\frac{1}{2} \varsigma(\gamma) \left(\frac{\partial^2 \Phi}{\partial x^2} + \frac{\partial^2 \Phi}{\partial y^2} \right) \right] \mathrm{d}\Omega_d. \tag{5.10}$$

A general energy formulation is used to obtain both terms in Eq. (5.8), where A is related to the thermal resistance of the system, while B is associated with the electrical resistance of the conductor. Similar to the thermal-fluid design optimization in Sect. 5.1.3 that follows later in this chapter, w_1 and w_2 in Eq. (5.8) are weighting values that scale the respective thermal and electrical portions of the objective function.

Thus, the full electrothermal optimization problem may be stated as

Find	γ
Minimize	Eq. (5.8)
Subject to	Eqs. (5.1) – (5.4)
	$\gamma = \tilde{\mathsf{H}}(\tilde{\psi})$
	$-R_f^2 \nabla^2 \tilde{\psi} + \tilde{\psi} = \psi$
	$-1 \le \psi \le 1$
	$\int\limits_{\Omega_d} \gamma \mathrm{d}\Omega_d - v_u \le 0$
Given	Eqs. (5.5) – (5.7).

Optimization Model and Results

The assumed 2-D model geometry and boundary conditions for the electrothermal design optimization of a three-terminal device are shown in Fig. 5.1. For simplicity, the design domain was assumed to be 1×1 m with a current source applied to Terminal 1 and electrical ground assumed for both Terminals 2 and 3. The external boundaries of the design domain were considered adiabatic, and design-dependent convective heat transfer was assumed inside the domain, as described above in relation to Eq. (5.7).

In this example, the solid material state was assumed to take thermal and electrical conductivity values of $k_o = 200$ W/(m K) and $\varsigma_o = 3.5 \times 10^7$ S/m, respectively. A current of 2,000 A was assumed at Terminal 1, and a surface convection coefficient, $h_o = 5$ W/(m^2 K), corresponding to natural convection, per [46], was additionally assumed with an ambient reference temperature of $T_o = 293$ K (i.e., 20 °C).

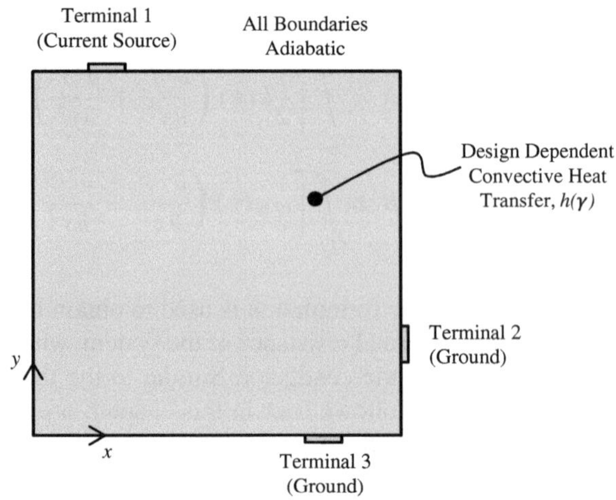

Fig. 5.1 Assumed 2-D simulation domain for electrothermal optimization example. A current source is assumed at Terminal 1 with electrical ground at Terminals 2 and 3. Design-dependent convective heat transfer with $h(\gamma)$ per Eq. (5.7) is assumed inside the domain with all remaining external boundaries considered adiabatic

The design domain was meshed with approximately 8,790 triangular finite elements having ~35,600 DOF, and the optimization problem was solved using commercial finite element solvers [15] coupled with a MMA optimizer [90]. Values of $\gamma_l = 0.45$, $\gamma_u = 0.75$, and $\varepsilon_b = 5 \times 10^{-3}$ were used in Eq. (5.7). A Helmholtz filtering routine with a filter radius of $R_f = 5 \times 10^{-3}$ was also implemented to enforce a minimum design length scale, as described in Sect. 4.1.

As an initial trial, weighting values of $w_1 = 0$ and $w_2 = 1$ in Eqs. (5.8)–(5.10) were assumed along with a 20 % conductor volume fraction constraint ($v_u = 0.2\,\text{m}^2$) to solve the pure electrical resistance minimization problem. The optimal solution was found in 50 iterations and took less than 5 minutes of computational time on a two-core laptop with 2.50 GHz processors and 3 GB of RAM. The bus bar design solution is shown on the left in Fig. 5.2 including normalized temperature contours and normalized total current density vectors. The surface convection heat transfer coefficient distribution is shown on the right in Fig. 5.2. Observe that a bus bar with the minimum length possible is constructed between the current source (Terminal 1) and the nearest ground (Terminal 2) to minimize the electrical resistance of the structure. Furthermore, Fig. 5.2 highlights the manner in which the surface heat transfer coefficient follows the conductor surface boundary with a zero value inside the solid material and a two-order of magnitude lower value outside of the bus bar, as specified per Eq. (5.7).

Extending the above results, two additional objective function weighting value pairs were examined to visualize the trade-off between the electrical and thermal terms in F_o, Eqs. (5.8)–(5.10). The bus bar design results for w_1 and w_2 pair values corresponding to Case I = [0, 1], Case II = [0.35, 0.65], and Case III = [0.5, 0.5] are shown in Fig. 5.3 on the left, center, and right, respectively.

The calculated bus bar electrical resistance, R, in terms of the voltage drop, V, and electrical current, i, is

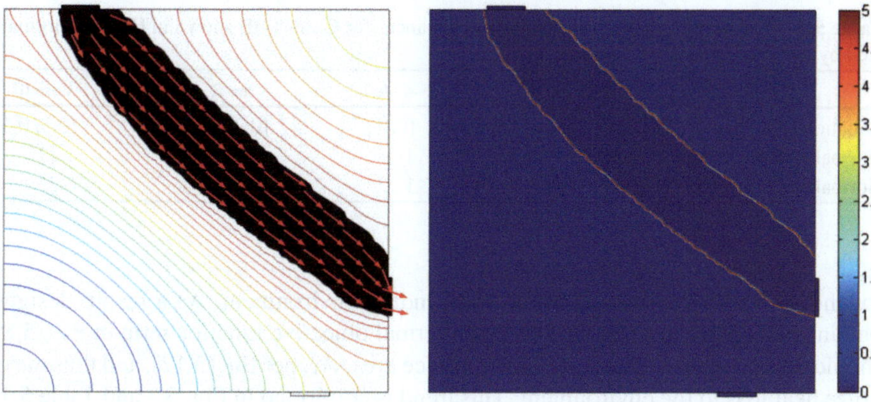

Fig. 5.2 Optimal bus bar design for 20 % conductor volume fraction and minimum electrical resistance, $[w_1 = 0, w_2 = 1]$, with normalized temperature contours and total current density vectors (*red arrows*) shown on the *left*. *Note* the *red colored temperature contours* indicate the highest temperature; *dark regions* = solid material; *light regions* = void. The corresponding final heat transfer coefficient distribution on the domain is shown on the right in units of W/(m^2 K)

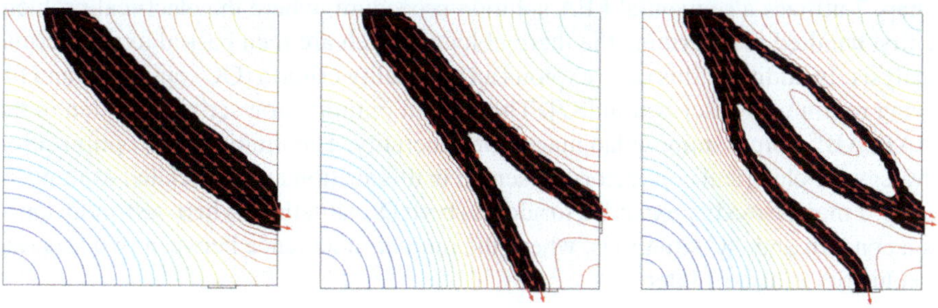

Fig. 5.3 Optimal bus bar designs for 20 % conductor volume fraction and weighting value (i.e., w_1 and w_2) pairs of [0, 1], [0.35, 0.65], and [0.5, 0.5] are shown, respectively, on the *left*, *center*, and *right* with normalized temperature contours and total current density vectors (*red arrows*). *Note* the *red colored temperature contours* indicate the highest temperature; *dark regions* = solid material; *light regions* = void

$$R = \frac{V}{i}, \tag{5.11}$$

while convective thermal resistance, $R_{\mathrm{th(cnv)}}$ per [45], is expressed in terms of the applied heat transfer coefficient, h, and conductor surface area, A_s, as

$$R_{\mathrm{th(cnv)}} = \frac{1}{hA_s}. \tag{5.12}$$

These two metrics are provided in Table 5.1 for each design shown in Fig. 5.3. Convective thermal resistance is used as a metric here since the sole heat loss mechanism

Table 5.1 Electrical and convective thermal resistances for Cases I–III shown in Fig. 5.3 (normalized by the respective Case I values)

	Case I	Case II	Case III
Equation (5.8) $[w_1, w_2]$ values	[0, 1]	[0.35, 0.65]	[0.5, 0.5]
Normalized electrical resistance	1	1.05	1.13
Normalized convective thermal resistance	1	0.48	0.33

for the system is convection to the surrounding environment. As a result, designs obtained using greater weighting of the thermal objective function term in Eq. (5.8) should demonstrate greater conductor surface area, A_s, per Eq. (5.12), and thus maximize heat loss to the environment. This trend is confirmed in Fig. 5.3 and Table 5.1, where designs that prioritize heat transfer exhibit greater branching and lower convective thermal resistance at the expense of thin electrical conductors having greater electrical resistance.

A numerical example similar to the above study is provided in Chap. 7, for reference, along with sample code for solving the representative problem. The code in Chap. 7 utilizes a sequential FEA solution procedure, where the electrical problem is first handled to determine the thermal loads which are then passed to the thermal analysis. In addition, this code exploits an optimality criteria (OC) optimizer and filtering routine, as described in [87]. The custom optimization script is written in the MATLAB®[1] programming language, for generality. The multiphysics optimization strategy for electrical conductors presented in this section and the related example in Chap. 7 may be readily extended to address both DC situations where self-inductance is a concern and AC situations, where the mutual inductance between two separate conductors is also of interest.

5.1.2 Design Optimization of Thermal-Structural Systems

Many electromechanical applications involve components that dissipate heat, and as a result, undergo structural deformation and thermally induced stress. One such application is found in the power control unit of modern hybrid and electric vehicles. The power module is a key power conversion component in these vehicles. It electrically connects the battery and driving motor(s), and provides multiple functions including driving motor power, regenerating power, boosting operational voltage, and charging the battery. Due to these multiple functions, the electronic devices found in power modules are exposed to high power density and experience high operational temperatures upwards of 423 K (150 °C). These high temperatures impose significant reliability challenges for the associated electronics package and often lead to critical thermally induced stresses. While inverter/converter modules are

[1] MATLAB® is a registered trademark of The MathWorks, Inc.

common in today's hybrid vehicles, they are also found in a variety of other sustainable engineering applications that involve energy generation and storage including wind turbines, solar energy installations, and future wireless charging systems. In all of these cases, multiphysics topology optimization has the potential to play a key role in performance enhancement.

Substrate damage is a typical end result of high-temperature-induced stress found in power electronics packaging. Direct bond copper substrates have been used in recent power module designs due to superior thermal performance. A DBC substrate consists of an insulating ceramic layer sandwiched between two copper layers. Often, additional materials may be deposited on top of the copper layers. The different materials have distinct thermal properties including different CTEs. Copper has a CTE that is 3–5 times larger than those of ceramic materials found in power module substrates. The CTE mismatch results in significant thermally induced stress at the interfaces between the electronics device and the substrate. Therefore, for reliable power modules, it is critical to minimize the thermally induced stress in the die and the corresponding bonding layer. Various efforts related to fabrication and material design have been devoted to solving such thermal stress problems. Stress reduction may be achieved by parametrically varying the thickness of the ceramic or metal layers. However, beyond simple material changes or thickness modifications, a new approach to thermal stress reduction and delamination prevention is to modify the overall substrate dimensions (e.g., size) and introducing a step layer at the edges of the metal layer; see for example [72]. This interesting approach suggests the use of numerical optimization methods in the resolution of thermal stress and delamination issues in power module substrates.

In this section, a parametric study of a sample electronics package is presented, where the thickness of the top and bottom metal layers of the DBC substrate is varied in order to minimize the stress in the substrate to the device bonding layer. While bonding layer stress reduction may be achieved via asymmetric DBC metal layer thickness, this reduction comes at the cost of increased complexity in terms of the base fabrication of the DBC structure. Instead, level set-based topology optimization is applied to the design of the top copper layer of a DBC substrate, per [76], having symmetric metal layer thickness in order to better minimize the thermally induced stress that occurs at the bonding layer between the electronics device and the top metallic layer of the DBC. Here, we adopt a 2-D pattern design approach instead of a cross-sectional configuration design strategy to avoid additional changes to the fabrication process. Thus, following the method introduced in Sect. 4.1.1, a 2-D level set field is first prepared for the design variables, and then the level set contour is projected into a 3-D space to construct a model for finite element analysis. The optimization code was implemented following the level set topology optimization approach with geometric reinitialization scheme from [93]. Finally, a 2-D design is directly obtained, which is utilized as a photo mask pattern for fabrication of a prototype DBC.

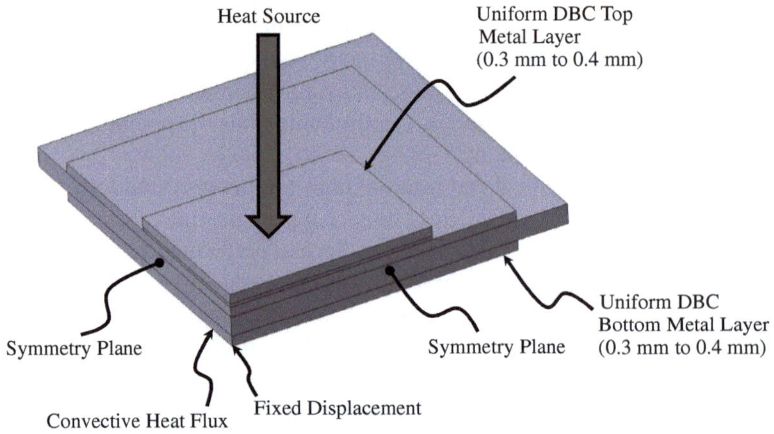

Fig. 5.4 One-quarter symmetry solid model geometry of electronics package used for parametric sizing study

5.1.2.1 Parametric Size Analysis of a Planar Electronics Package

The deformation and resulting thermal stress state of a laminated composite such as the electronics package described above is highly dependent on the CTE of the individual layers of the composite structure as well as the thickness of each lamina. Here, this electronics package is studied initially via a parametric sizing analysis to explore this stress-to-layer-thickness relationship.

The model geometry utilized in this study is shown in Fig. 5.4, where a one-quarter symmetry model is adopted. Standard material properties were assumed for the various material layers, which moving from bottom to top include copper (in-plane ~17 mm square, thickness ~0.3 mm), aluminum nitride (in-plane ~21 mm square, thickness ~0.6 mm), copper (in-plane ~17 mm square, thickness ~0.3 mm), solder (in-plane ~9 mm × 12 mm, thickness ~0.1 mm), and silicon (in-plane ~9 mm ×12 mm, thickness ~0.3 mm). A heat transfer coefficient of $h = 2{,}000$ W/(m^2 K) at a reference temperature of $T_o = 293$ K was applied to the bottom of the package. A device power of 50 W was applied to the silicon layer.

In this investigation, the top metal layer plus the bottom metal layer of the DBC are each varied in size from 0.3 to 0.4 mm in 0.01 mm increments to determine the optimal configuration that minimizes the maximum Von Mises stress in the package solder bond layer. Thus, a total of 121 simulations were run in parametric fashion over 9 h using an eight-core workstation with 2 GHz processors and 64 GB of RAM to determine the response of the composite structure for all possible layer thickness combinations. The computational mesh settings adopted for the parametric study involved an approximate total number of DOF ranging from 3.2×10^5 to 4.2×10^5 depending on the structural configuration (i.e., layer thickness geometry) under consideration.

The deformed state of the package is shown in Fig. 5.5, where the DBC top and bottom metal layers have thicknesses of 0.3 and 0.4 mm, respectively. The structural response for the 121 different possible layer thickness combinations was computed, and configuration #11 shown in Fig. 5.5 produces the lowest maximum Von Mises stress, 52.1 MPa, in the solder bond layer with a maximum device temperature of 118.8 °C. The response surface from the full parametric study is shown in Fig. 5.6, where we see that asymmetric package designs comprising a DBC with a top metal layer that is thinner than the bottom metal layer consistently perform better. Nonetheless, such asymmetric DBC designs may pose potential problems in terms of deformation due to residual stresses during high temperature substrate fabrication, and a package design strategy that retains the symmetry of the DBC metal layer thickness is desirable.

Optimization Model and Results

From Sect. 3.3, the governing equations for the steady-state thermal stress analysis of a structure may be written as

$$-\nabla \cdot \left[k(\gamma)\,\nabla T\right] = Q, \tag{5.13}$$

$$\mathbf{F}_\alpha = \mathbf{B}_e \mathbf{C}(\gamma)\alpha(\gamma)(T - T_o), \tag{5.14}$$

$$-\nabla \cdot \sigma = \mathbf{F}_\alpha, \tag{5.15}$$

where T is the temperature state variable, and σ is the matrix of thermally induced stresses. The thermal conductivity, $k(\gamma)$, stiffness tensor, $\mathbf{C}(\gamma)$, and coefficient of thermal expansion, $\alpha(\gamma)$, are all design-dependent material physical parameters. Note that T_o in the coupling expression, Eq. (5.14), is a reference temperature required for the calculation of the resultant thermal strains.

For the thermal stress design optimization problem, the design-dependent material physical parameters are interpolated as follows:

$$E(\gamma) = E_o + (E_s - E_o)\tilde{\mathsf{H}}(\gamma), \tag{5.16}$$

$$k(\gamma) = k_o + (k_s - k_o)\tilde{\mathsf{H}}(\gamma), \tag{5.17}$$

$$\alpha(\gamma) = \alpha_o + (\alpha_s - \alpha_o)\tilde{\mathsf{H}}(\gamma). \tag{5.18}$$

In all three of the above expressions, the subscript, o, indicates a reference value, which in this case, corresponds to void versus the subscript, s, which corresponds to a solid material physical parameter. Note that the stiffness tensor, $\mathbf{C}(\gamma)$ in Eq. 5.14, depends on the design-dependent elastic modulus, $E(\gamma)$, in Eq. 5.16. Additionally, $\tilde{\mathsf{H}}(\gamma)$ is the relaxed (differentiable) Heaviside function, as described in Sect. 4.1.

A metal layer pattern for the DBC structure is designed by minimizing the objective function,

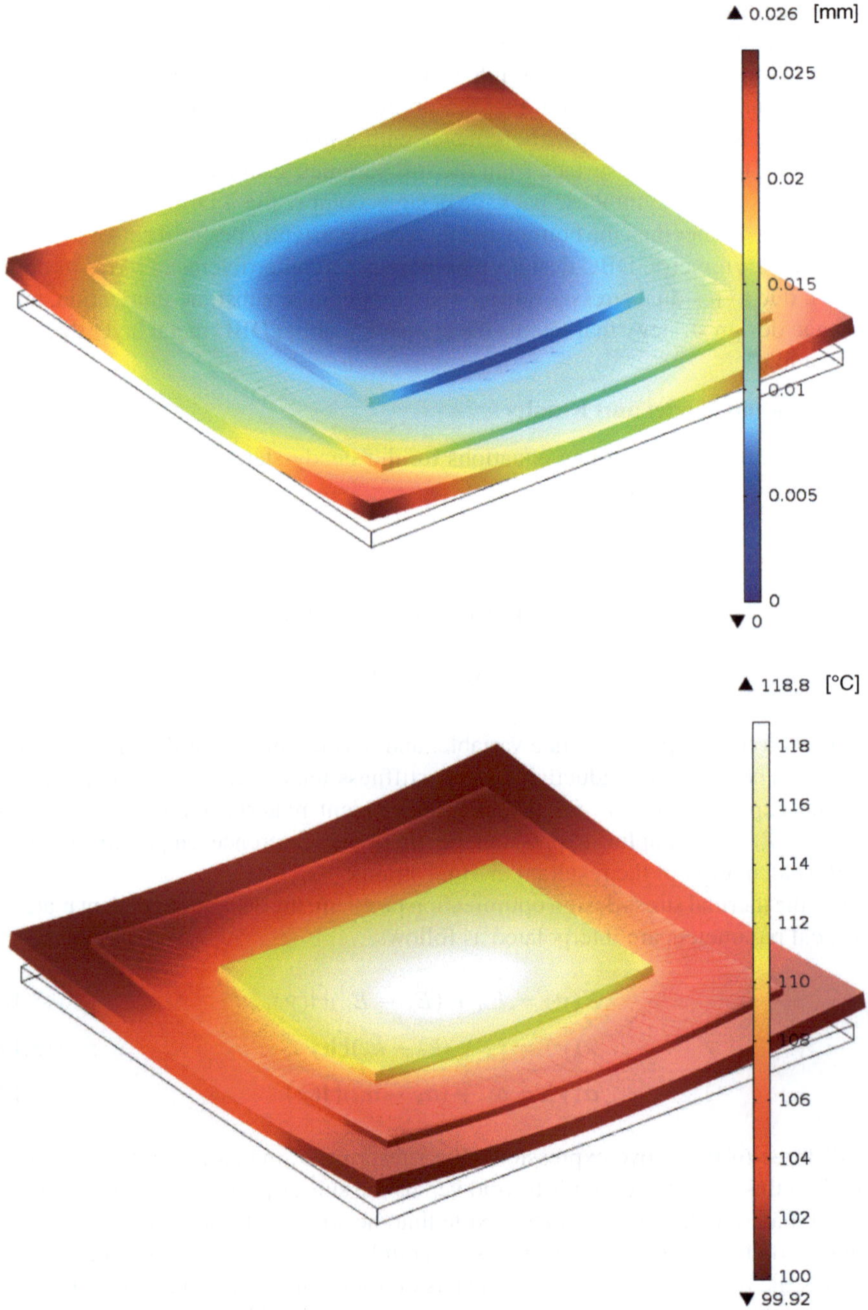

Fig. 5.5 Representative deformed package geometry with superimposed total displacement contours (*upper image*) and temperature contours (*lower image*) for parametric case #11 with the DBC top metal layer thickness equal to 0.3 mm and the DBC bottom metal layer thickness set to 0.4 mm

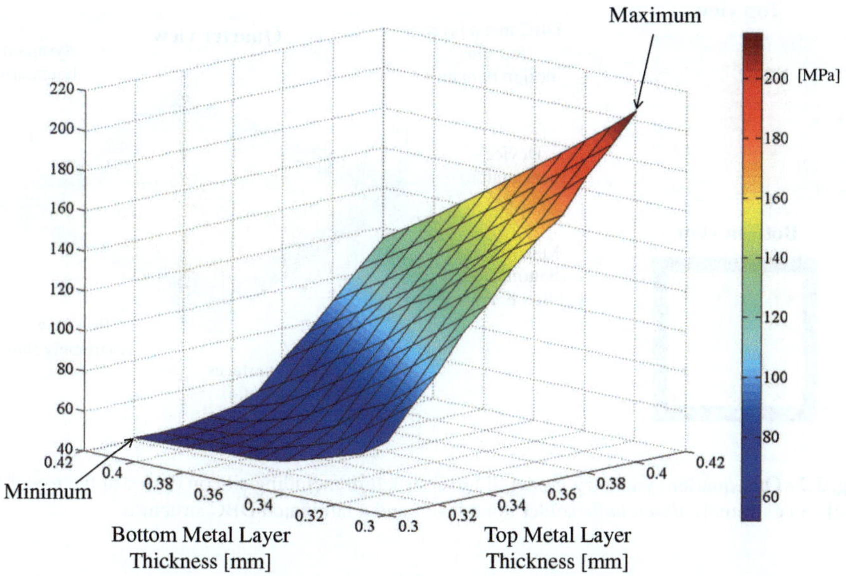

Fig. 5.6 Response surface from the parametric size analysis for the DBC with *top* and *bottom* metal layers of varying thickness ranging from 0.3 to 0.4 mm in 0.01 mm increments for a total of 121 possible configurations

$$\mathsf{F}_o = \int_0^{\varepsilon} \sigma : \mathrm{d}\varepsilon; \quad \varepsilon = \frac{1}{2}\left[\nabla \mathbf{u} + (\nabla \mathbf{u})^{\mathrm{T}}\right], \tag{5.19}$$

which is defined on the bonding layer between the device and DBC. In Eq. (5.19), σ is the Cauchy stress tensor, ε is a vector of the resulting strains, and $\mathbf{u}$ is the displacement vector in the domain of interest.

A one-quarter symmetry model of an assumed multilayer electronics package including the device at top and a laminated DBC structure on bottom with a solder bond attachment layer positioned in between is shown in Fig. 5.7. Here, the top metal layer of the DBC structure is specifically designed to minimize the strain energy density, via Eq. (5.19), in the mechanical bonding layer between the DBC and device.

The numerical model for the optimization is similar to the one presented above; that is, the same device power dissipation, heat transfer convective flux, and boundary conditions were assumed. Additionally, the same material physical parameters were used for each layer in the numerical model. In this case, the top metal layer was meshed in 2-D using 1,764 quadrilateral elements, while the subsequent full 3-D structure was meshed using $\sim 45 \times 10^3$ hexahedral elements. The optimal solution was found using commercial finite element software [15] in 100 iterations and took

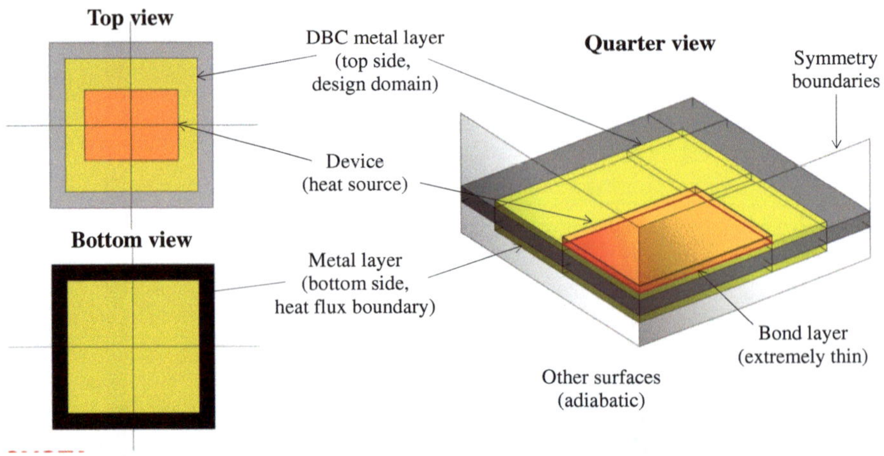

Fig. 5.7 One-quarter symmetry model of DBC structure including silicon device at the *top* of the stack, an extremely *thin middle* solder bond layer, and a laminated DBC structure

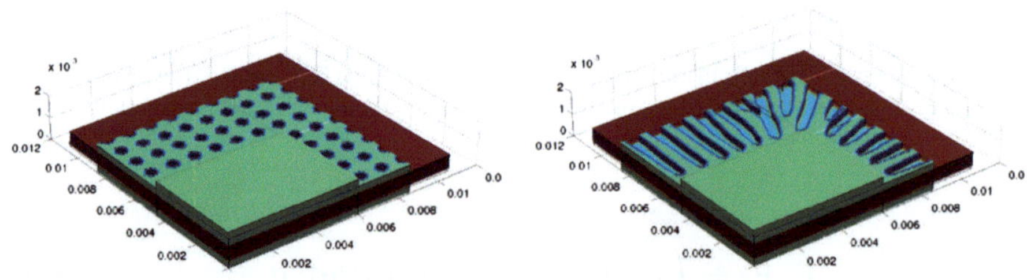

Fig. 5.8 Package geometry assumed for initialization of the level set algorithm (*left*) and final structure including optimized *top* metal layer of DBC (*right*)

approximately 2.5 h on a 12-core workstation with 2.40 GHz processors and 24 GB of RAM.

The geometry assumed to initialize the level set optimization algorithm is shown in Fig. 5.8 on the left along with the final optimized structural geometry on the right. A finger-like structure is obtained which acts to minimize the strain energy at the corners of the solder bond layer (i.e., the region of maximum stress). Compared with a structure having a fully dense top metal layer (refer to Fig. 5.7), the maximum Von Mises stress is reduced by approximately 28 % for the optimized top metal layer; refer to the image on the right in Fig. 5.8.

An advantage of this approach is that the 2-D level set result may be directly implemented as the photo resistor pattern in a standard DBC batch etching, dicing, and polishing process. Thus, standard techniques may be exploited in the fabrication of complex DBC structures; see for example Fig. 5.9.

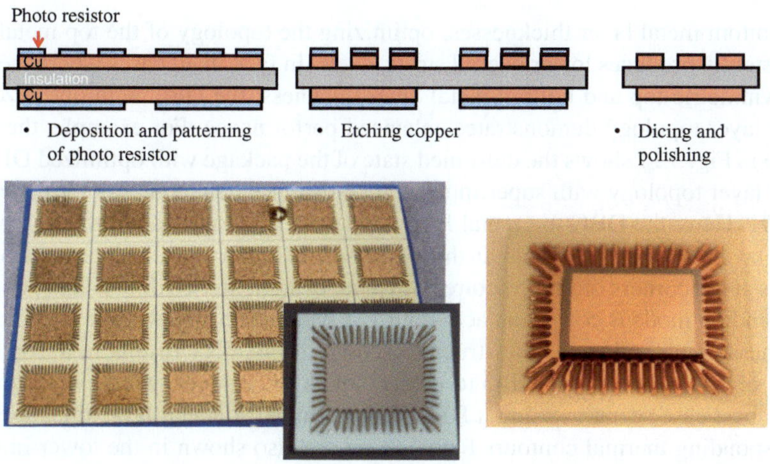

Fig. 5.9 Prototype DBC structures batch fabricated using the level set optimization result with a standard DBC etching process, as shown schematically at the *top* of the figure

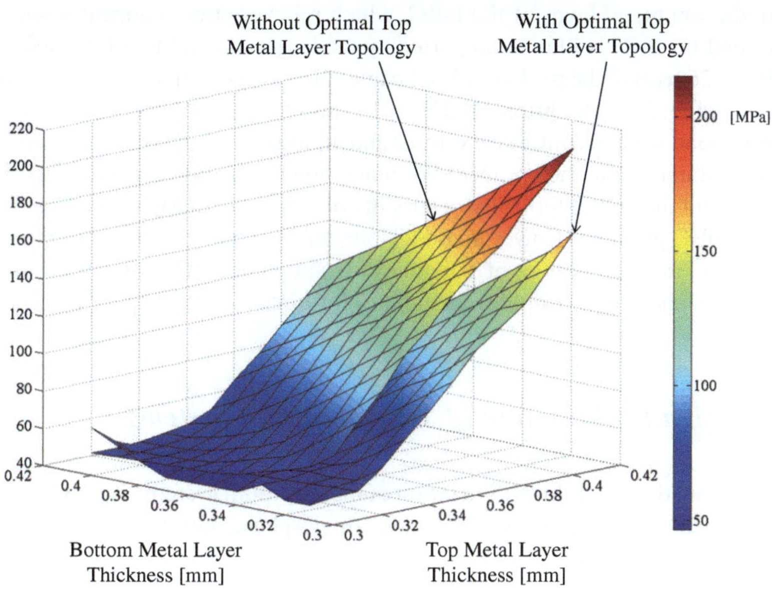

Fig. 5.10 Overlaid response surfaces from the parameter studies for the DBC with and without optimal topology *top* metal layer

Discussion

As a follow-up to the earlier parametric sizing study, a similar study was performed using the optimized DBC top metal layer topology, and the response surfaces from the two studies are plotted together in Fig. 5.10. For the same range of DBC top and bottom metal layer thicknesses, optimizing the topology of the top metal layer consistently produces lower overall stress levels. In fact, in all cases where the DBC has symmetric top and bottom metal layer thickness, the DBC with optimized top metal layer topology demonstrates enhanced performance. For example, the upper image in Fig. 5.11 shows the deformed state of the package with optimized DBC top metal layer topology with superimposed total displacement contours for parametric case #1. Here, the DBC top metal layer thickness is 0.3 mm, and the DBC bottom metal layer thickness is 0.3 mm. In this case, the maximum total displacement, 32 μm, occurs at the corners of the structure with plate bending in two dimensions; note that this bending mode is consistent across all 121 parametric models. The specific layer thickness combination for the structure shown in Fig. 5.11 results in a Von Mises stress of 54.7 MPa in the DBC to device solder bond layer, which is within 5 % of the best case scenario using a DBC with asymmetric metal layer thickness. The corresponding thermal contours for this case are also shown in the lower image of Fig. 5.11, for reference, and the maximum device temperature is 120.2 °C.

A physical explanation for the enhanced performance of the optimized DBC is that the finger-like structure introduces greater compliance and establishes an equal stress state in the top metal layer of the DBC, which relieves stress concentrations in the solder bond layer. This idea is supported by the larger maximum total displacement (i.e., 32 vs. 26 μm) of the package plus the overall package Von Mises stress contours, as shown in Fig. 5.12. Additionally, Fig. 5.13 shows the volumetric strain contours at the solder bond to device interface surface and highlights lower strain at the corners of this critical interface in the optimized package (upper image) versus the well-known stress concentration that typically occurs in corners of regularly shaped rectangular or square DBC packages (lower image). Note that the color contours in Fig. 5.13 are plotted using the same range of volumetric strain values (with higher strain levels shown in red and lower strain levels in blue) for direct comparison of the images.

5.1.3 Design Optimization of Thermal-Fluid Systems

In this section, the topology optimization of thermal-fluid systems is presented in the context of heat sink design for electronics. Single-phase liquid (i.e., incompressible fluid) cooled systems are considered, although logical extensions of the method include air cooling. Single-phase liquid cooled microchannel heat sink technologies are of interest due to their potential for handling very high electronic device heat fluxes [45, 88]. Comprehensive research has shown that traditional straight channel systems may exhibit very low thermal resistance at the cost of relatively high pressure drop [37]. Alternatively, novel manifold heat sink designs have been proposed to reduce the pressure drop penalty while maintaining some of the heat transfer benefits of straight microchannels [40, 92].

Accordingly, this example study is focused on the topology optimization of a nonstraight branching channel heat sink arranged in a manifold-like configuration. Optimal coolant channel paths are found by minimizing a two-term objective

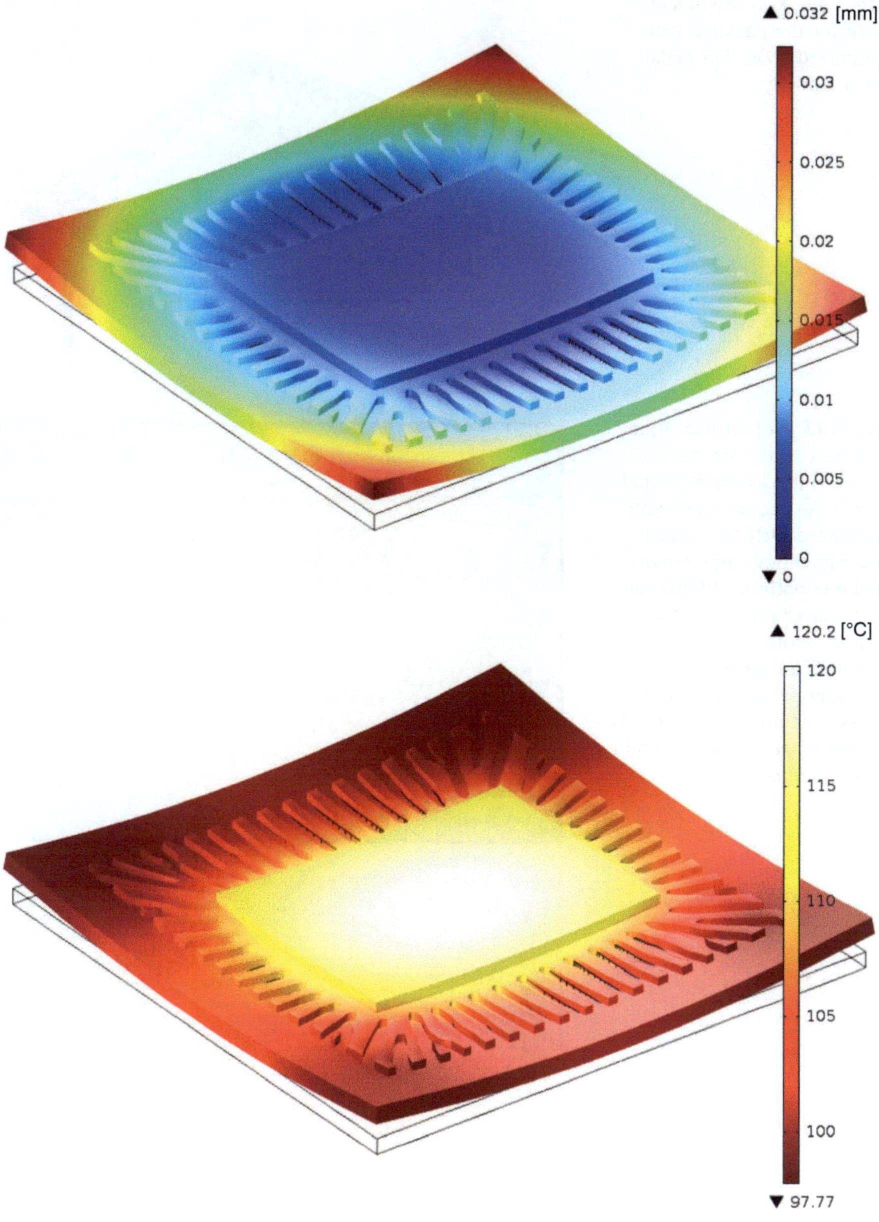

Fig. 5.11 Representative deformed package geometry with superimposed total displacement contours (*upper image*) and temperature contours (*lower image*) for parametric case #1 with the DBC *top* metal layer thickness equal to 0.3 mm and the DBC *bottom* metal layer thickness set to 0.3 mm

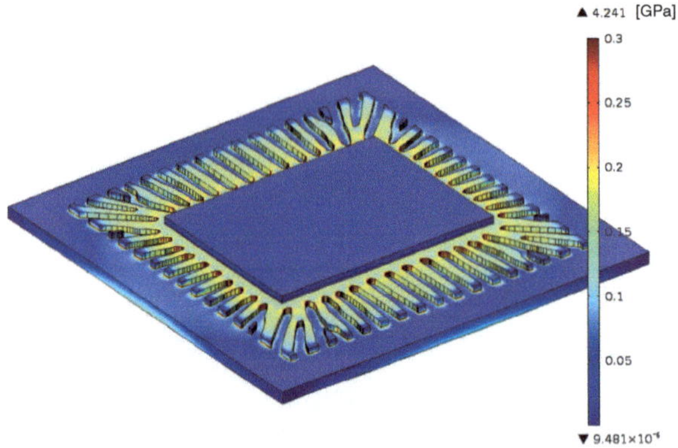

Fig. 5.12 Von Mises stress state for the package with optimized DBC *top* metal layer topology

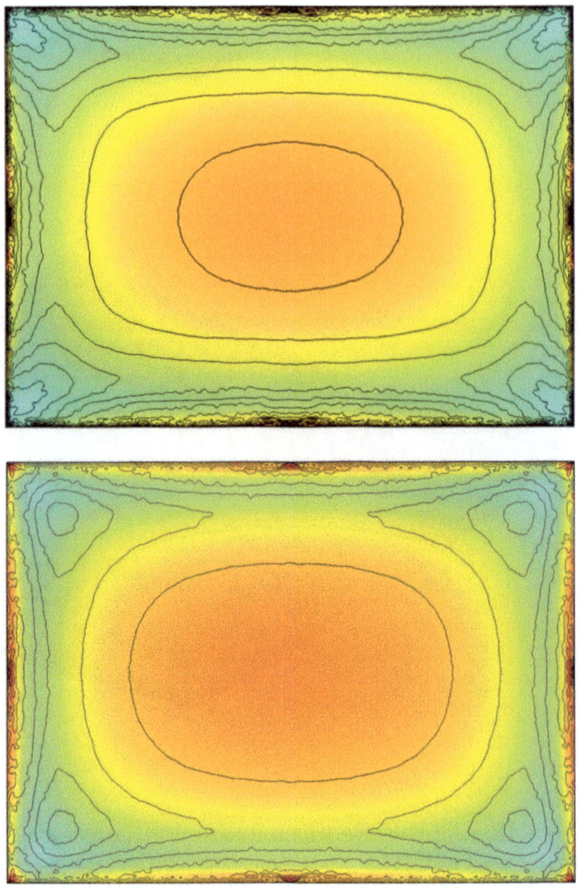

Fig. 5.13 Volumetric strain contours at the interface between the solder bond and device for the package with optimized DBC *top* metal layer topology (*upper image*) and non-optimized DBC *top* metal layer topology (*lower image*). Note that the color contours are plotted using the same volumetric strain range with higher strain levels shown in *red* and lower strain levels in *blue*

function. The first term is related to the domain average temperature, while the second term is associated with the fluid power dissipated (or flow resistance) in the domain. Following the material distribution approach outlined in Chap. 4, the effective inverse permeability and thermal conductivity of an assumed porous medium are specified as a function of the material design variable, γ; see [8, 95] for additional details. Related methods have previously been used in the design optimization of ducts [77], heat transfer surfaces [20, 44], and microfluidic mixers [4].

Following [19, 24, 77, 83], and similar to the single physics fluid flow example from Sect. 4.1, the governing equations for steady-state Navier-Stokes flow in an idealized porous medium are given as

$$\nabla \cdot \mathbf{v} = 0, \quad \text{and} \tag{5.20}$$

$$\rho\left(\mathbf{v} \cdot \nabla \mathbf{v}\right) = -\nabla P + \nabla \cdot \left\{\eta\left[\nabla \mathbf{v} + \left(\nabla \mathbf{v}\right)^{\mathrm{T}}\right]\right\} - \tilde{\alpha}\left(\gamma\right)\mathbf{v}. \tag{5.21}$$

Equation (5.20) represents the fluid incompressibility constraint, while Eq. (5.21) describes laminar fluid flow. In these expressions, ρ and η are again the fluid density and dynamic viscosity, respectively. The inverse permeability of the porous medium, $\tilde{\alpha}(\gamma)$, is assumed to approximately represent an actual porous medium, per [77]. The state variables once more include the fluid pressure, P, and velocity field terms in the vector, $\mathbf{v}$.

In addition to fluid flow, the governing equation for steady-state convection-diffusion heat transfer is

$$\rho C_p\left(\mathbf{v} \cdot \nabla T\right) = \nabla \cdot \left[k\left(\gamma\right)\nabla T\right] + Q, \tag{5.22}$$

where C_p represents the heat capacity, $k(\gamma)$ is the thermal conductivity of the fluid, and Q is the volumetric power density.

Following the approach in Sect. 5.1.1, a multi-term objective function, F_o, was implemented to optimize for both heat transfer and fluid flow. Specifically, the objective was specified to minimize the mean temperature and total fluid power dissipated in the system,

$$\mathsf{F}_o = w_1 A + w_2 B, \quad \text{where} \tag{5.23}$$

$$A = \int_{\Omega_d} \left\{k\left(\gamma\right)\left(\nabla T\right)^2 + \rho C_p\left[T\left(\mathbf{v} \cdot \nabla T\right)\right]\right\} \mathrm{d}\Omega_d, \quad \text{and} \tag{5.24}$$

$$B = \int_{\Omega_d} \left[\frac{1}{2}\eta \sum_{i,j}\left(\frac{\partial \mathrm{v}_i}{\partial x_j} + \frac{\partial \mathrm{v}_j}{\partial x_i}\right)^2 + \sum_i \tilde{\alpha}\left(\gamma\right)\mathrm{v}_i^2\right] \mathrm{d}\Omega_d. \tag{5.25}$$

As described in [19, 24], The terms w_1 and w_2 in Eq. (5.23) are user selected weighting values that scale the respective thermal and fluid portions of the objective function. 'Tuning' of these values assists in convergence and modifies the resulting optimal topology by affecting the dominance of one physical process relative to another. For simplicity, these weighting values were selected manually in the numerical example that follows, although an adaptive scaling strategy for automatic determination of the weighting of the individual objective function terms could alternatively be adopted for this problem; more information is provided later in Sect. 5.4.1.

Focusing on the optimal steady-state fluid flow and channel layout, the thermal conductivity and inverse permeability of the porous medium were, respectively, interpolated using a penalty method and convex interpolation scheme; refer to Chap. 4. These effective properties, k and $\tilde{\alpha}$, were interpolated via the main design parameter, γ, which varied from 0 (minimally porous, lower thermally conductive solid) to 1 (high thermally conductive fluid), respectively, as follows:

$$k(\gamma) = (k_{max} - k_{min})\,\gamma^{p} + k_{min}, \quad \text{and} \tag{5.26}$$

$$\tilde{\alpha}(\gamma) = \tilde{\alpha}_{min} + (\tilde{\alpha}_{max} - \tilde{\alpha}_{min})\,\frac{q(1-\gamma)}{q+\gamma}. \tag{5.27}$$

where the penalization power, p, is set to 3 and the tuning parameter, q, is set equal to 0.1; see Sect. 4.1 for additional details and [24] for more information regarding the specific problem setup and assumptions.

Thus, the full optimization problem may be formulated as

Find	γ
Minimize	Eq. (5.23)
Subject to	Eqs. (5.20)–(5.22)
	$\int_{\Omega_d} \gamma \, d\Omega_d - v_u \leq 0$
	$0 \leq \gamma \leq 1$
Given	Eqs. (5.26) and (5.27).

Optimization Model and Results

The model described in this section is adapted from [24], where a very thin heated plate was assumed. A related 2-D model is used for computational efficiency since a typical microchannel cooling system has in-plane dimensions that greatly exceed its out-of-plane thickness; note that the effect of the fin height on heat sink performance is examined in the parametric size analysis presented in the next section.

The implemented 2-D model is shown in Fig. 5.14 along with applied boundary conditions and loads. A laminar flow (i.e., Re $<$ 2100), fixed temperature inlet velocity condition was specified along with a zero pressure, convective flux outlet condition. A uniform heat flux was applied to the design domain. The optimal topology for this model was computed using commercial finite element software [15] with

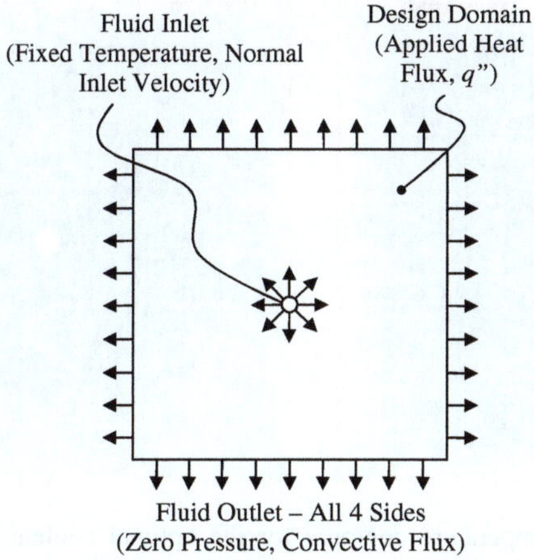

Fig. 5.14 Assumed 2-D model with loads and boundary conditions (the model represents a thin square plate with a center fluid inlet and fluid outlets along all four sides. The design domain is subject to uniform heat generation). Reprinted from [24, Fig. 5a], Copyright (2012), with permission from American Society of Mechanical Engineers (ASME)

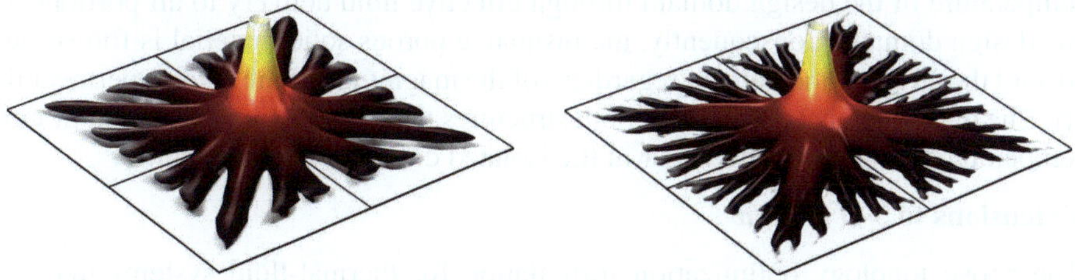

Fig. 5.15 Optimal cooling channel topologies for the 2-D model from Fig. 5.14 obtained using coarse (*left*) and fine (*right*) computational meshes. The *color contours* indicate the normalized fluid velocity magnitude with larger velocity at higher elevations

both fine and coarse meshes having approximately 2,900 and 25,000 DOF, respectively, to examine the effect of mesh refinement on the final topology since a filtering routine was not implemented. The refined mesh result required approximately 1.4 h computational time on a dual-core laptop with 2.5 GHz processors and 3 GB of RAM.

The optimal channel topologies for the model described above are shown in Fig. 5.15, where the coarse mesh result is on the left and the fine mesh result is on the right [24]. A 40 % fluid volume fraction was used in both cases. Observe that a branching channel structure is obtained, and greater mesh refinement produces a higher level of branching complexity.

Figure 5.16 shows the normalized temperature contour results for both mesh conditions. Note that the ratio of the weightings for the thermal to fluid portions of the objective function is approximately 30:1; refer to [24] for further details. These

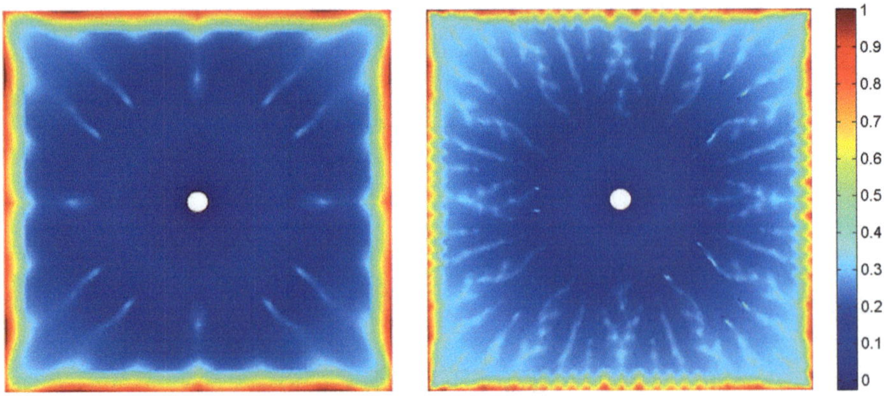

Fig. 5.16 Normalized temperature contours for the optimal cooling channel topologies from Fig. 5.15 using coarse (*left*) and fine (*right*) computational meshes. Reprinted from [24, Figs. 9a, b], Copyright (2012), with permission from American Society of Mechanical Engineers (ASME)

results illustrate the manner in which the branching channel structure decreases the temperature of the design domain through effective fluid delivery to all portions of the design domain. Consequently, the insulative porous solid material is forced out toward the edges of the domain regardless of the magnitude of the inverse permeability. The main diagonal branches of both structures then work to reduce the maximum temperature, which tends to occur in the isolated corners of the domain.

Extensions to 3-D Design

The above topology optimization formulation for thermal-fluid systems may be extended to handle 3-D laminar flow conjugate heat transfer problems with increased computational cost. In the following brief example, a similar square thin plate (width to thickness ratio = 10:1) optimization domain is considered with a center, fixed temperature and velocity (laminar flow) fluid inlet and four, zero pressure side fluid outlets. A heat flux was applied to the bottom wall of the domain, and a material interpolation strategy similar to [95] was utilized. Here, solid regions are assigned a greater thermal conductivity value than fluid regions (e.g., copper and water, respectively) to better capture the 3-D fin effect of the solid portion of the heat exchange structure.

A Pareto front for the design optimization problem is shown in Fig. 5.17, where the normalized fluid temperature contours are superimposed on an isosurface of the optimal cooling channel flow path with a slice through the solid domain (i.e., dark regions). In this figure, design results with greater weighting of the thermal portion of the objective function are found higher up along the ordinate axis. In contrast, the design results with greater weighting of the fluid flow resistance term of the objective function in Eq. (5.23) are shown further to the right along the abscissa. As a point

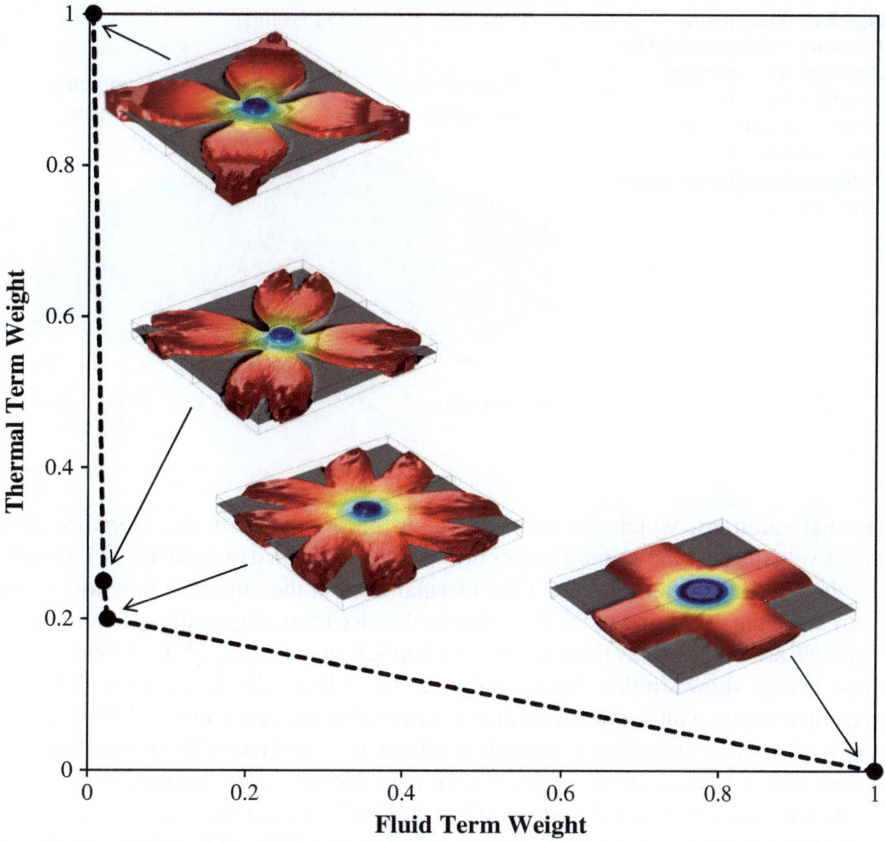

Fig. 5.17 Pareto front for 3-D topology optimization of a *thin plate* heated from *below* and having a *center*, fixed temperature and velocity fluid inlet with four, zero pressure side outlets. Normalized temperature contours are superimposed on an isosurface of the optimal cooling channel flow path with a slice through the solid domain (i.e., *dark regions*)

of reference, each solution shown in Fig. 5.17 was again obtained using commercial software [15] with a computational mesh having ~480,000 DOF and required approximately 4 hr computational time on a six-core workstation with 3.46 GHz processors and 48 GB of RAM.

Observe that a simple cross flow structure that minimizes the distance from the fluid inlet to outlets is obtained when priority is placed on minimizing the fluid power dissipated in the domain; refer to the lower right solution in Fig. 5.17. In contrast, increasing the priority of the thermal term in the objective function (i.e., minimizing the mean temperature of the domain, as well) produces branching cooling channel topologies, as seen in the middle two images shown in Fig. 5.17. These cooling

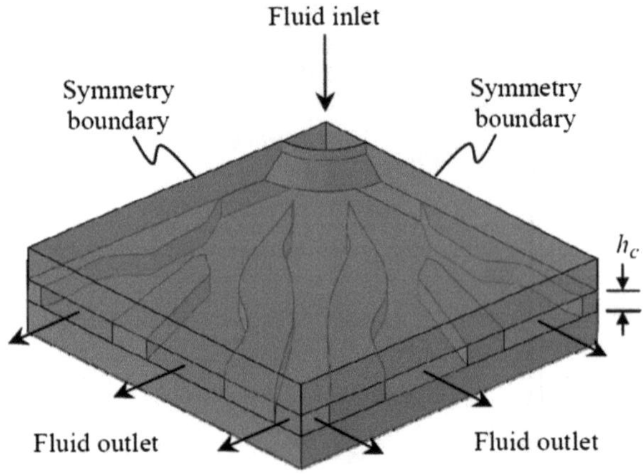

Fig. 5.18 One-quarter symmetry solid model CAD geometry for branching channel cold plate (*arrows* indicate fluid flow direction; h_c represents the channel height parameter)

channel solutions, which are very similar to those studied in the literature [89], effectively manage the domain corner hot spots, as explained in detail in [24]. Finally, by placing the most emphasis on the thermal term in the objective function, while decreasing the weighting of the fluid flow resistance term, diagonally extending fluid channels are obtained, as illustrated in the top left image in Fig. 5.17. Interestingly, these results show similar characteristics to the 2-D results described in [24] and presented above, which indicates that, for certain classes of problems, a 2-D modeling approach may be sufficient in providing effective, manufacturable design solutions in less time due to reduced computational cost. However, the fluid expansion (i.e., a 3-D effect) seen toward the edges of the design domain in the images in Fig. 5.17 indicates that for design domains of larger thickness, a fully 3-D method is likely the best approach to capturing the relevant structural topology.

5.1.3.1 Parametric Size Analysis of a Finned Heat Sink

In this section, the derived 2-D cooling channel topology for a square design domain presented above is used to demonstrate the utility of parametric size analysis techniques for heat sink design; the study is based on the work in [21]. Here, the effect of channel aspect ratio on the performance of the branching channel cold plate design is examined, and the resulting trade-off between heat transfer and pumping power is explained.

The coarse mesh topology optimization result (on the left in Fig. 5.15) was used in the development of the CAD geometry for a cold plate structure, as illustrated in Fig. 5.18, where a one-quarter symmetry model is assumed for computational efficiency. The model geometry consists of a jet plate (shown transparent) with a center nozzle that directs the coolant downwards into the channel structure, through the radial branching channels, and then out through the sides of the cold plate.

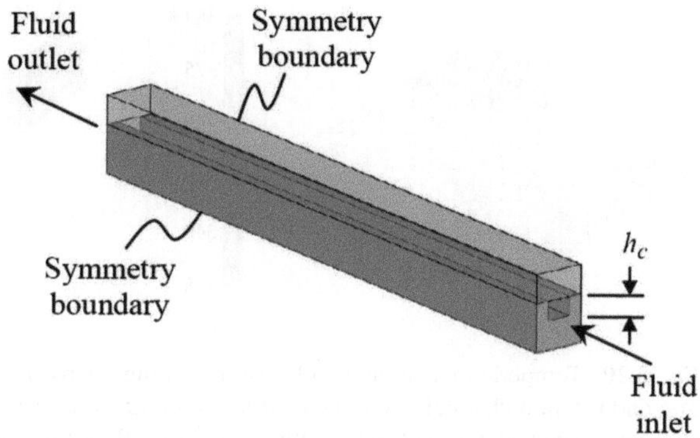

Fig. 5.19 Symmetry solid model CAD geometry for a straight microchannel cold plate (*arrows* indicate fluid flow direction; h_c represents the channel height parameter)

The coarse mesh topology optimization result from Sect. 5.1.3 was employed due to the relatively small overall size of the cold plate and to reduce the overall complexity of the microscale channel system. The size of the full plate is 17.2 mm square with a 1.26 mm initial thickness. The initial channel height parameter, h_c, value was set to 0.5 mm and was swept from this initial value up to 2.0 mm in 0.5 mm increments using the 'parametric sweep' feature in a commercially available software package. The cold plate channel width, w_c, ranged from approximately 0.66 to 2.42 mm.

For comparison purposes, a straight microchannel cold plate was also studied. A symmetry model of a single microchannel representing a 17.2 mm square cold plate with an initial thickness 1.26 mm is shown in Fig. 5.19. The channel width was set equal to the minimum channel width of the branching cold plate design, i.e., 0.66 mm, while the channel height was again swept from 0.5 to 2 mm in 0.5 mm increments.

In both cases, the channel aspect ratio was assumed as the channel height divided by the minimum channel width, $\zeta = h_c/w_{c\_min}$. Therefore, four separate values of the channel aspect ratio (~0.75, 1.5, 2.25, and 3) were considered for each of the described parametric computational models.

Parametric Model and Results

For each of the finned cold plate channel structures, shown in Figs. 5.18 and 5.19, a thermally conductive heat sink material was assumed having a thermal conductivity, specific heat capacity, and density of 160 W/(m K), 900 J/(kg K), and 2700 kg/m^3, respectively. The coolant fluid thermal conductivity, specific heat capacity, density, and dynamic viscosity were, respectively, set to 0.654 W/(m K), 4182 J/(kg K), 982 kg/m^3, and 4.4E-4 Pa s. An insulative material was also assumed for the jet plate and straight microchannel cap with a thermal conductivity of 0.26 W/(m K), a specific heat capacity of 1700 J/(kg K), and a density of 1150 kg/m^3.

Appropriate symmetry boundary conditions were applied to each of the models described in Figs. 5.18 and 5.19. Additionally, a uniform heat flux, $q'' = 100$ W/cm^2, was applied to the bottom side of each channel structure. The fluid inlet was set to an

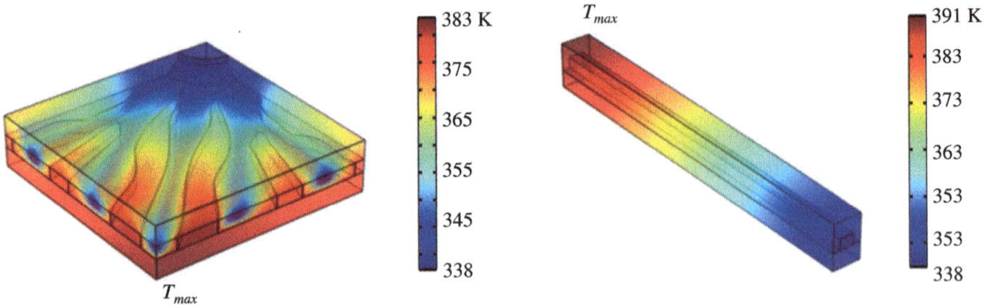

Fig. 5.20 Temperature contour results for branching microchannel cold plate with 0.15 L/min flow rate and 0.5 mm channel height (*shown transparent for clarity on left*). Temperature contours for straight microchannel cold plate with 0.15 L/min flow rate and 0.5 mm channel height (*shown transparent for clarity on right*)

elevated temperature above 323 K. A zero pressure, convective flux condition was set at each fluid outlet boundary.

An automated parameter study was set up using commercial finite element software [16] to sweep the height parameter, h_c, in each model from 0.5 to 2 mm, as previously described. The computational models were solved at each channel height parameter step over a range of fluid inlet volumetric flow rates spanning 0.025–0.15 L/min. For each case, appropriate mesh refinement was selected with greater element density in the fluid channels and at the channel walls.

The temperature contour results for the branching cold plate with a 0.5 mm channel height at maximum inlet fluid flow rate are shown in Fig. 5.20 on the left. Likewise, the results for the straight channel structure are shown in Fig. 5.20 on the right. Observe that the maximum temperature, T_{max}, for the branching channel design, 383 K, occurs at the outside corners of the square plate. For the straight channel design the maximum temperature, $T_{\text{max}} = 391$ K, occurs at the end of the channel after the thermal boundary layer has developed.

Two performance metrics are used to evaluate the effectiveness of each of the cold plate designs over the various height parameters and flow rate ranges. The unit thermal resistance, R''_{th}, and pressure drop, ΔP, are calculated for each cold plate as

$$R''_{\text{th}} = \frac{T_{\text{max}} - T_{\text{in}}}{q''}, \quad \text{and} \tag{5.28}$$

$$\Delta P = P_{\text{in}} - P_{\text{out}}. \tag{5.29}$$

In Eq. (5.28), T_{max} is the maximum cold plate temperature, T_{in} is the fluid inlet temperature, and q'' is the applied heat flux. The pressure drop in Eq. (5.29) is calculated as the difference between the cold plate inlet pressure, P_{in}, and outlet pressure, P_{out}.

Figure 5.21 (left side image) provides the unit thermal resistance as a function of flow rate for both cold plates over all channel height parameter values. The unit thermal resistance of the branching microchannel cold plate for $h_c = 0.5$ mm is less

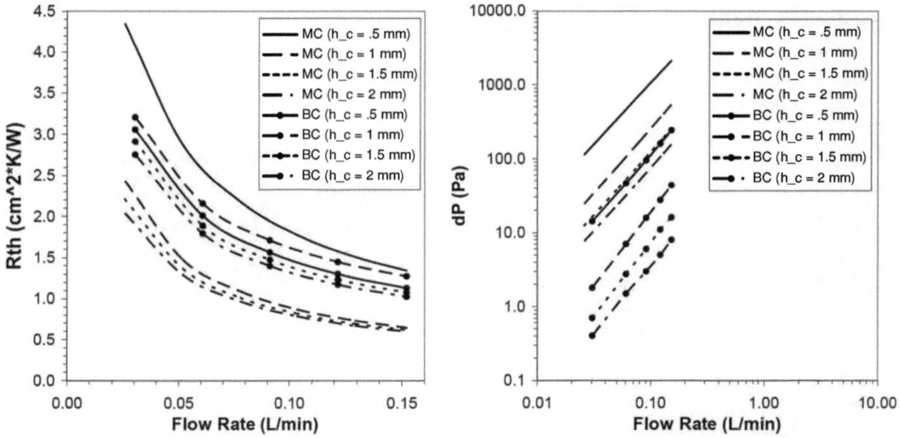

Fig. 5.21 Comparison of unit thermal resistance (*left*) and pressure drop (*right*) for branching (BC) and straight microchannel (MC) cold plates

than that for the straight microchannel system. However, as the height and channel aspect ratio is increased, the unit thermal resistance of the straight microchannel system drops to values that are approximately one-half of those for the branching microchannel system. Additionally, the branching channel cold plate unit thermal resistance increases as channel height is increased from 0.5 to 1 mm. However, as the channel height is further increased to 1.5 and 2 mm, the unit thermal resistance then progressively decreases.

The pressure drop as a function of flow rate for both cold plate designs is also shown in Fig. 5.21 (right side image). Observe that at the 0.5 mm channel height the branching cold plate design exhibits a maximum pressure drop value of 0.24 kPa compared with 2.1 kPa for the straight microchannel design. This trend in magnitude is consistent across the various channel aspect ratios where the branching microchannel cold plate generally exhibits one to two orders of magnitude lower pressure drop.

While the straight microchannel system outperforms the branching channel structure in terms of heat transfer at larger channel aspect ratios, the considerably larger ΔP may be prohibitive in applications where pumping power is limited. This characteristic of microchannels is well known, and it is a primary obstacle to widespread application [37].

5.1.3.2 Extensions to Prototype Design

The combined use of topology optimization with parametric design studies may lead to interesting solutions to well-established engineering problems in terms of the development of novel prototypes and products [25, 26, 28]. As a first example, the heat transfer performance of this branching channel cold plate may be increased by

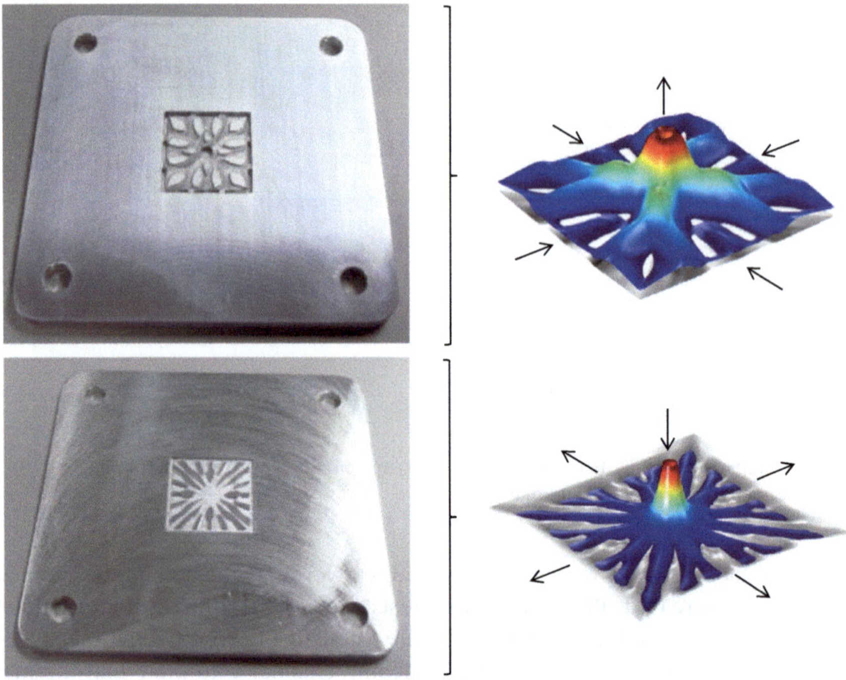

Fig. 5.22 Layers 1 and 2 of a multi-pass branching microchannel cold plate. Layer 1 is shown on the *bottom row* with the corresponding flow channel topology optimization result on the *right*. Layer 2 is shown on the *top row* with the corresponding flow channel topology optimization result on the *right*. *Arrows* indicate the fluid flow direction; *color contours* indicate the normalized fluid velocity value with larger values at higher elevations

considering a multi-pass structure, as discussed in [24, 27], where two layers of a heat exchange structure were designed using a topology optimization approach and coupled together to further lower the cold plate unit thermal resistance. The local microchannel regions of the first two heat transfer layers (i.e., Layers 1 and 2) of this design are illustrated in Fig. 5.22 and show the optimized channel and fin structures. A cross-section image of the final diffusion bonded heat sink is shown in Fig. 5.23, and the thermal-fluid performance of this cooler is discussed in [27].

A second example prototype cold plate design is also shown in Fig. 5.24 on the right, where the optimal channel topology (shown on left) was found for a uniformly heated rectangular design domain with a center jet fluid inlet and two side outlets. This cold plate design was integrated into a manifold hierarchical microchannel heat sink arrangement and was experimentally studied in [23]. The conclusion from this investigation was that the optimized cooler provided increased heat transfer through efficient flow distribution with essentially the same pressure drop as a similar cooler that utilized jet impingement of a uniformly heated flat plate.

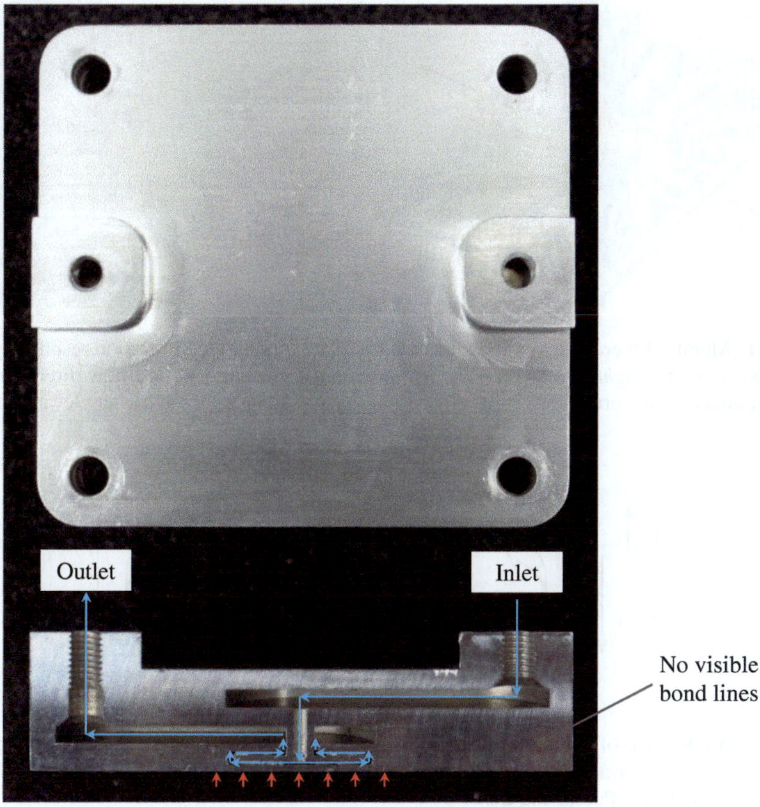

Fig. 5.23 *Top view* of fabricated multi-pass branching microchannel heat sink with cross-section view shown below. The cross-section view is shown with the fluid flow path (*blue arrows*) and heat input region (*red arrows*) overlaid. The multilayer cold plate design is fabricated out of aluminum material using a diffusion bonding process to create a continuous thermal path through the thickness of the assembly. The cross-section image highlights the fact that there are no visible *bond lines* post-assembly

Thus, the integrated optimization, analysis, and design procedure may be exploited in conjunction with the fluid manifold design procedure outlined Chap. 4 in the development of a variety of efficient real-world heat transfer and fluid flow structures. It should be noted that while conjugate heat transfer considering forced convection and laminar flow was the main focus of this section, the formulation may be readily extended to address additional effects such as natural convection, where the Boussinesq approximation is applied to couple the fluid temperature and velocity fields [2]. Furthermore, the topic of thermal-fluid topology optimization for higher Re number turbulent flows is also a developing research area [54] that will further increase the potential of using such methods for realistic cold plate, heat exchanger, and cooling jacket design.

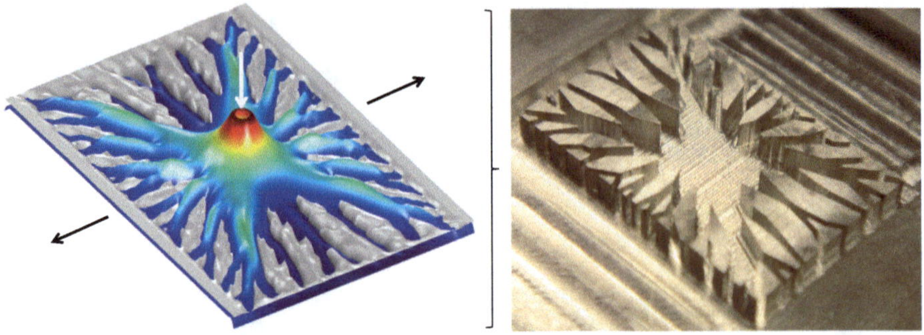

Fig. 5.24 Manifold hierarchical microchannel cold plate topology optimization results (*left*) and prototype structure (*right*). *Arrows on the left side image* indicate the fluid flow direction; *color contours* indicate the normalized fluid velocity value with larger values at higher elevations

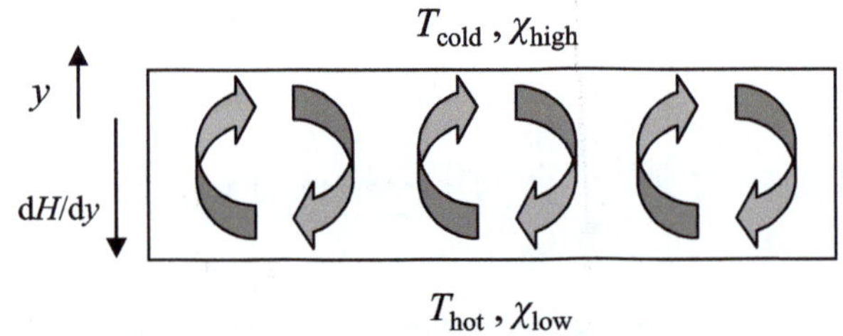

Fig. 5.25 A schematic of a classic closed-loop magnetically controlled convective heat transfer system illustrating the thermomagnetic siphoning effect. Reprinted with permission from [59, Fig. 1], Copyright (2012), AIP Publishing LLC

5.1.4 Design Optimization of Thermomagnetic Convective Systems

While cold plate microchannel design for electronics applications was considered in Sect. 5.1.3, a related concept is presented in this section, where the control of fluid flow through the use of thermomagnetic field effects for heat transfer purposes is considered. The use of stationary or time-varying magnetic fields to control the motion of magnetically susceptible fluids is a well-known phenomenon [10, 36, 69, 80, 84, 85], and the use of this phenomenon represents an interesting concept for the thermal management of electronic systems. Here, an optimal magnetic field distribution is shown to enhance thermomagnetic siphoning and consequently provide increased heat transfer for localized heat sources. While implementation and assuring correct fluid properties for maximum thermomagnetic siphoning effect [50] is still challenging, exploiting such physical couplings for the passive pumping of fluid may offer advantages in future heat transfer devices for electromechanical systems.

A schematic of a classic magnetically controlled convective heat transfer system is shown in Fig. 5.25. In this system, a magnetic fluid container is subjected to a

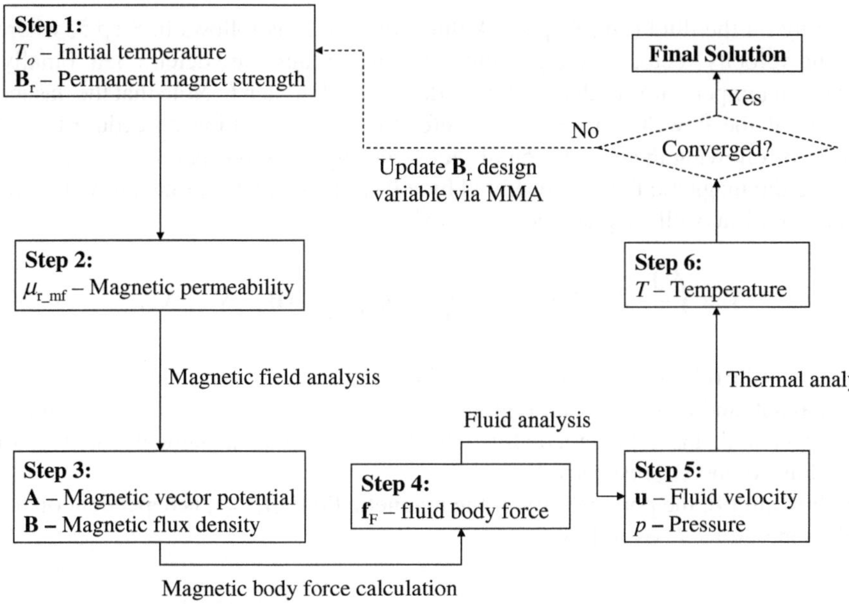

Fig. 5.26 Flowchart of computations for the multiphysics analysis of a magnetically controlled convective heat transfer system. *Note* the *dashed portions* indicate optimization steps. Reprinted from [60, Fig. 2], Copyright (2012), with permission from American Society of Mechanical Engineers (ASME)

temperature gradient, where the top side of the system is fixed at a cold temperature, T_{cold}, and the bottom side is held at a hot temperature, T_{hot}. Additionally, a magnetic field gradient, $-\mathrm{d}H/\mathrm{d}y$, is assumed along the y-axis, as shown. As discussed in [84], the magnetic susceptibility, χ, of the fluid is inversely proportional to the temperature. Thus, cold fluid at the top of the container is more strongly magnetized and drawn to the region of higher magnetic field strength thus displacing the hot fluid at the bottom of the container. This unique system involves three physics, including magnetic fields generated either by permanent magnets (PMs) or electromagnets in air and external to the magnetic fluid container, fluid flow emanating from the associated magnetic fluid body force, and heat transfer related to the aforementioned fluid flow.

The flowchart for the multiphysics steady-state computational analysis of the system just described is shown in Fig. 5.26. In this work, it is assumed that the magnetic field gradient is generated by a PM positioned in an air domain outside of the magnetic fluid container. Thus, the non-linear analysis procedure begins with Step 1, where the temperature gradient (or heat source plus heat sink) is specified along with the magnet residual magnetic flux density, $\mathbf{B}_r$, and an initial condition system temperature, T_o. These inputs determine the magnetic fluid permeability, μ_{r_mf}, via the temperature-dependent fluid susceptibility. After Step 2, a magnetic field analysis is performed to compute the magnetic vector potential, **A**, and flux density, **B**, (i.e., Step 3) which allows for the calculation of the magnetic body force,

$\mathbf{f}_F$, acting on the fluid (i.e., Step 4). A fluid flow analysis follows in Step 5, where the fluid velocity vector, $\mathbf{v}$, and pressure, P, state variables are determined. Finally, the system temperature, T, distribution is determined in Step 6. Note that the dashed portions of the flowchart in Fig. 5.26 refer to the optimization procedure that is explained following the description of the governing equations below.

Since the magnetic field is generated by a PM, there is no external current in the system, and Maxwell's equations reduce to[2]

$$\nabla \times \left(\frac{1}{\mu_r} \nabla \times \mathbf{A}\right) = \nabla \times \left(\frac{1}{\mu_r} \mathbf{B}_r\right); \qquad (\mathbf{B} = \nabla \times \mathbf{A}), \tag{5.30}$$

where μ_r is the relative magnetic permeability (including the surrounding air, permanent magnet, and magnetic fluid regions of the analysis domain), $\mathbf{A}$ is the magnetic vector potential, and the magnetic flux density and residual magnetic flux density of the PM are $\mathbf{B}$ and $\mathbf{B}_r$, respectively.

Following [9], the permeability of the magnetic fluid, μ_{r_mf}, is dependent on the fluid temperature, T, as follows:

$$\mu_{r\_mf} = \mu_o (1 + \chi), \quad \text{with} \quad \chi = 4\pi \left(\frac{C \cdot \rho}{T \cdot M_w}\right), \tag{5.31}$$

where χ is the fluid magnetic susceptibility, C is the Curie constant of the magnetic fluid, ρ is the fluid density, and M_w is the molecular weight.

Using Eq. (5.31), the magnetic fluid body force is then calculated as

$$\mathbf{f}_F = \frac{1}{2} \mu_o \chi \nabla \left(H^2\right), \quad \text{with} \quad H = \left|\frac{1}{\mu_{r\_mf}} \nabla \times \mathbf{A}\right|, \tag{5.32}$$

where μ_o is the permeability of free space, and H is the magnitude of the applied magnetic field. This body force term is then utilized in the fluid flow analysis for which the Navier–Stokes equation and standard incompressibility constraint are assumed.

$$\rho (\mathbf{v} \cdot \nabla \mathbf{v}) = -\nabla P + \nabla \cdot \left\{\eta \left[\nabla \mathbf{v} + (\nabla \mathbf{v})^{\mathrm{T}}\right]\right\} + \mathbf{f}_F, \quad \text{and} \tag{5.33}$$

$$\nabla \cdot \mathbf{v} = 0. \tag{5.34}$$

In the above expressions, the magnetic fluid dynamic viscosity is given as η, while $\mathbf{v}$ is the fluid velocity vector, and P is the pressure state variable.

In terms of heat transfer within the system, the standard convection-diffusion equation is assumed inside the fluid domain

$$\rho C_p (\mathbf{v} \cdot \nabla T) = \nabla \cdot (k \nabla T) + Q, \tag{5.35}$$

[2] Reprinted, with permission, from [59], Copyright (2012), AIP Publishing LLC.

where k and C_p are the thermal conductivity and specific heat capacity of the magnetic fluid, respectively, T is the temperature, and Q is the applied volumetric power density. For any solid domain, Eq. (5.35) reduces to

$$0 = \nabla \cdot (k_s \nabla T) + Q_s, \tag{5.36}$$

where the subscript, s, indicates material physical parameters and thermal loads associated with the solid regions of the system [83].

The multiphysics analysis of the system is performed by solving Eqs. (5.30)–(5.36) using the finite element method. Using the finite element formulation, Eqs. (5.30)–(5.36) can be represented in matrix form as

$$\mathbf{K}_M(\mathbf{T}) \cdot \mathbf{A} = \mathbf{F}_M, \tag{5.37}$$

$$\mathbf{K}_F(\mathbf{v}) \cdot (\mathbf{v} \quad \mathbf{P})^{\mathrm{T}} = \mathbf{F}_F(\mathbf{A}), \tag{5.38}$$

$$\mathbf{K}_T(\mathbf{v}) \cdot \mathbf{T} = \mathbf{F}_{\mathrm{T}}, \tag{5.39}$$

where $\mathbf{K}_M$, $\mathbf{K}_F$, and $\mathbf{K}_T$ are the global system stiffness matrices, and $\mathbf{F}_M$, $\mathbf{F}_F$, and $\mathbf{F}_T$ are the corresponding global force vectors for the magnetic, fluid, and thermal equations, respectively. The coupled non-linear equations, Eqs. (5.37)–(5.39), are then solved numerically using commercial finite element software [15].

As described in [59], two design variables, γ_1 and γ_2, are assigned to each finite element in the design domain for the optimization problem. The x and y direction components of the residual flux density, $\mathbf{B}_r$, of each element are defined and interpolated as

$$B_{r,x} = B_{r\_PM} \cdot \gamma_1^3 \cdot \sin\{720^\circ \cdot (\gamma_2 - 0.5)\}, \tag{5.40}$$

$$B_{r,y} = B_{r\_PM} \cdot \gamma_1^3 \cdot \cos\{720^\circ \cdot (\gamma_2 - 0.5)\}. \tag{5.41}$$

In Eqs. (5.40)–(5.41), the first design variable, γ_1, controls the strength of the permanent magnet, B_{r_PM}. When $\gamma_1 = 0$ the corresponding finite element represents a void (i.e., air), and when γ_1 is equal to unity, the element represents a PM. The second design variable, γ_2, then controls the PM magnetization direction, where the design variable γ_2 is constrained between 0 and 1 (similar to γ_1) as is usual in topology optimization. When $\gamma_2 = 0.5$ in Eqs. (5.40)–(5.41), the angle of the magnetization direction relative to the y-axis is zero, and thus the magnetization direction points along the positive y-direction.

The optimal PM locations and magnetization directions that minimize the maximum system temperature, $T_{\max}$, of the system,

$$\mathsf{F}_o = T_{\max}, \tag{5.42}$$

are then found as per the procedure described in [59]. Thus, the full thermomagnetic convective optimization problem may be stated as

Find	γ_1 and γ_2
Minimize	Eq. (5.42)
Subject to	Eqs. (5.37) – (5.39)
	$0 \le \gamma_1 \le 1$
	$0 \le \gamma_2 \le 1$
Given	Eqs. (5.40) – (5.41).

In terms of implementation, the above formulated optimization problem is solved using an MMA optimizer [90], as shown in Fig. 5.26. Note that no volume constraint is applied to the design variables.

Multiphysics Analysis of Classic Thermomagnetic Convective System

To illustrate the general thermomagnetic instability effect [84], a 2-D system similar to that shown in Fig. 5.25 was studied. An image of the assumed numerical model is provided in Fig. 5.27, where a closed chamber sized 20 × 5 mm containing a magnetically susceptible fluid is subjected to a fixed top side temperature, T_{cold} = 293 K (20 °C), and a constant bottom side temperature, T_{hot} = 393 K (120 °C), with two permanent magnets sized 2.5 × 1 mm positioned in the surrounding air environment below the container. For this example, and the optimization example that follows, the magnetic fluid is assumed to have a thermal conductivity, density, dynamic viscosity, heat capacity, Curie constant, and molecular weight of 2.7 W/(m K), 1060 (kg/m^3), 2 × 10^{-3} Pa s, 3000 J/(kg K), 1 K m^3/mol, and 100 g/mol, respectively. In this specific example, each permanent magnet is assumed to have a strength of 0.2 T with magnetization in the *y*-direction.

The analysis domain was meshed using ~15,700 triangular elements for a total of approximately 44,400 DOF with greater element refinement near the walls of the fluid container. The solution of the non-linear multiphysics analysis (i.e., Steps 1 through 6 in Fig. 5.26) required roughly 1.5 minutes on a dual-core laptop computer with 3.0 GHz processors and 2 GB of RAM.

The simulation results for this model are shown in Fig. 5.28. Specifically, the magnetic vector potential field lines are shown in Fig. 5.28a, and from the magnetic field, the body force acting on the magnetic fluid is calculated using Eq. (5.32). The resulting magnetic fluid normalized body force distribution is shown in Fig. 5.28b. In the latter figure, the lighter colored regions near the PM material indicate a strong magnetic force. The fluid velocity contours with streamlines are shown in Fig. 5.28c indicating the formation of four recirculation zones with a maximum and average velocity of 1.95 × 10^{-2} m/s and 6.27 × 10^{-3} m/s, respectively, that draw cooler fluid

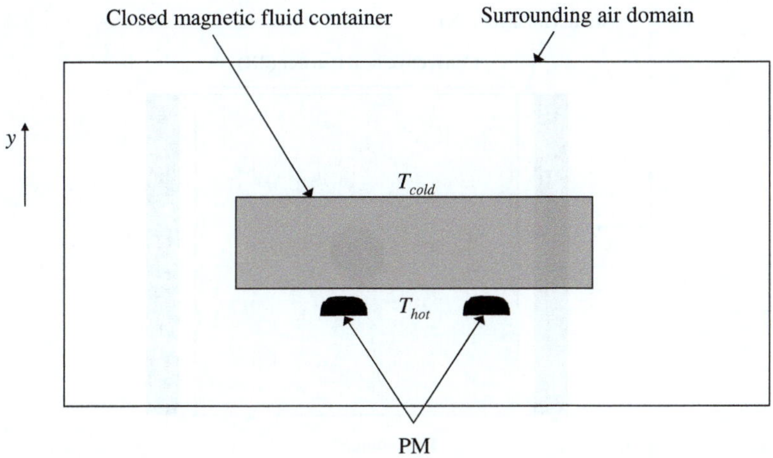

Fig. 5.27 Assumed 2-D simulation domain for a classic closed-loop magnetically controlled convective heat transfer system; refer to the corresponding conceptual schematic shown in Fig. 5.25

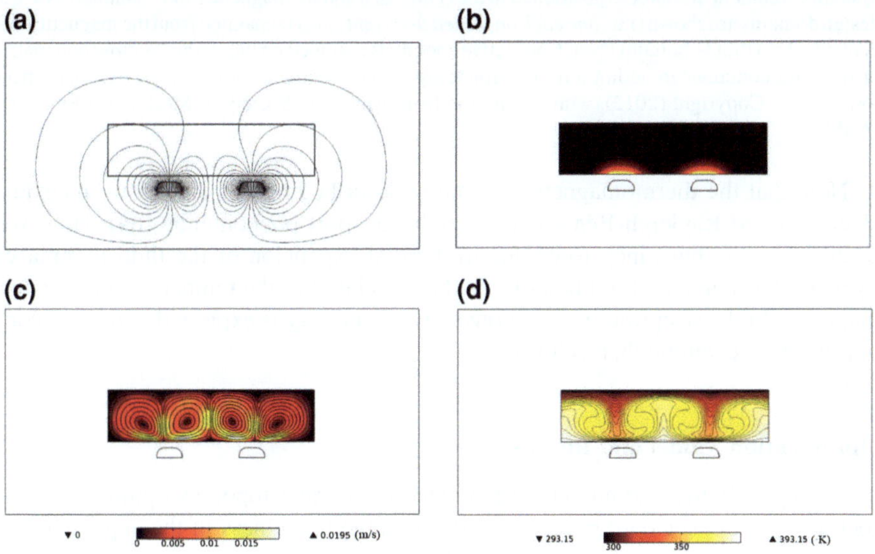

Fig. 5.28 Simulation results for the example system from Fig. 5.27: **a** magnetic vector potential field lines; **b** normalized magnitude of the magnetic fluid body force distribution (*note lighter regions* indicate larger magnitude); **c** magnetic fluid velocity contours with streamlines; and **d** temperature contours

from the top side down towards the bottom side of the container. The corresponding temperature contours are provided in Fig. 5.28d to illustrate the related convection effect.

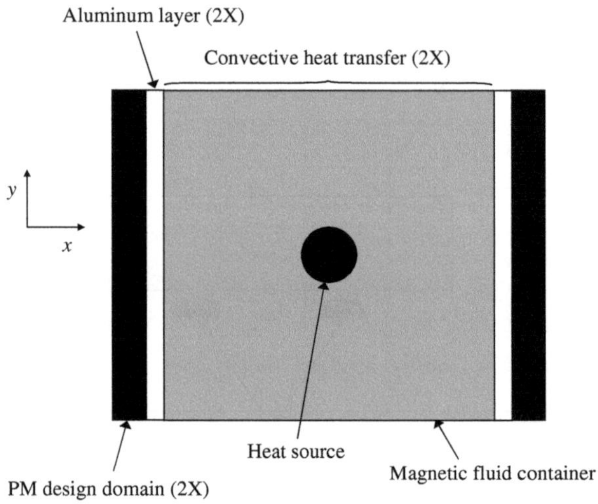

Fig. 5.29 Assumed 2-D simulation domain for magnetically controlled convective heat transfer system. Circular heat source is positioned in the *center* of a square magnetic fluid container. Two PM design domains are shown (i.e., *one each on the left and right sides*) separated from the magnetic fluid container by a thin aluminum layer. Convective cooling is assumed on the *top* and *bottom* boundaries of the fluid container in addition to a surrounding air environment (*not shown*). Reprinted from [60, Fig. 3], Copyright (2012), with permission from American Society of Mechanical Engineers (ASME)

Note that the thermomagnetic siphoning described here is similar to the extensively studied Rayleigh-Bénard convection transport phenomenon [84]. However, in this case the buoyancy force due to thermal expansion of the fluid is replaced by a similar magnetic fluid body force that is related to the temperature-dependent magnetic fluid susceptibility. This convective instability is exploited in the topology optimization example that follows.

Optimization Model and Results

Following [60], the next numerical example incorporates topology optimization and focuses on a nonstandard problem to illustrate the usefulness of the approach. The assumed geometry and key boundary conditions are given in Fig. 5.29, where a circular heat source of 625 W with a diameter of 4 mm is positioned in the center of a 20 mm square magnetic fluid container. Two 2×20 mm PM design domains are defined along the left and right edges of the magnetic fluid container, each separated from the fluid by a thin 1×20 mm aluminum layer representing the container wall. The aluminum material was assigned a thermal conductivity of 160 W/(m K), while the permanent magnet strength was set to 0.1 T. Additionally, an arbitrarily defined convective heat flux boundary condition of $180 \times (T\text{-}293)$ W/m^2 at a reference temperature of 293 K (20 °C) was defined along the top and bottom edges of the fluid container. Without fluid motion, the maximum temperature of the system is 449.6 K, for reference.

To mesh the analysis domain, approximately 6,240 quadratic elements with 82,900 DOF were used including a surrounding air region (not shown in Fig. 5.29) for the magnetic field analysis. It was found that higher element orders produced accurate solutions for the more complex fluid flow patterns being considered. Thus, the fluid velocity, $\mathbf{v}$, and pressure, P, terms were interpolated using third and second order finite elements, respectively. For the temperature, T, and magnetic vector potential, $\mathbf{A}$, second order finite elements were utilized. The solution of the full optimization problem required 450 iterations or approximately 12.5 h on a quad-core workstation with 3.0 GHz processors and 8 GB of RAM.

The optimal set of design variables, γ_1 and γ_2, is obtained by solving the previously described topology optimization problem, and the results are visualized in Fig. 5.30a. The darker regions in the design domain indicate PM material, while the light colored regions indicate air. Again, a volume constraint was not used to restrict the amount of the PM material, and the solution is almost entirely PM with the exception of a small air-gap magnetic flux barrier built in the middle of each design domain. The arrows in Fig. 5.30a also show the magnetization direction of the PM material. As can be seen in Fig. 5.30b, the optimal PM arrangement generates a strong magnetic field on the fluid side of the system, while minimizing the magnetic field in the surrounding air region on the opposite side; the resulting normalized magnetic fluid body force distribution is shown in Fig. 5.30c, for reference. This special arrangement of the PM magnetization direction, sometimes referred to as a Halbach array [39], is obtained as the optimal PM layout for the computed magnetic field.

The fluid velocity contours with streamlines are shown in Fig. 5.30d, where the average and maximum fluid velocities are, respectively, 3.37×10^{-3} m/s and 1.58×10^{-2} m/s. These values are more than four times larger in magnitude when compared with the best performer from non-optimized test cases. Additionally, a very clear four eddy flow pattern is evident. Observe that the PM in each quadrant of the system acts as a combined fluid 'pump' plus 'conveyor' to, respectively, draw the fluid in toward the PM and then move the fluid along the length of the PM and vertical sides of the container thus creating symmetric convective flow patterns that efficiently transport the heat from source to sink. This result is further visualized in Fig. 5.30e, where the maximum temperature of the system is 402 K (129 °C). Here, the heat is siphoned off of the cylindrical heat source and distributed along the width of each heat sink at the top and bottom of the magnetic fluid container.

For comparison purposes, several PM layouts were manually tested to check the effectiveness of the thermomagnetic convection phenomenon without optimization. The PM domains were split up into four blocks, and each quadrant domain was filled with a 0.1 T PM with various assumed magnetization directions. The number of the possible combinations is large even if the magnetization directions are constrained to be parallel to the Cartesian axes. So, for simplicity, only the four best performing designs obtained through a manual study are presented here. The selected magnetization directions are shown in Table 5.2, and the corresponding numerical results are shown in Table 5.3. The latter table provides the computed values for

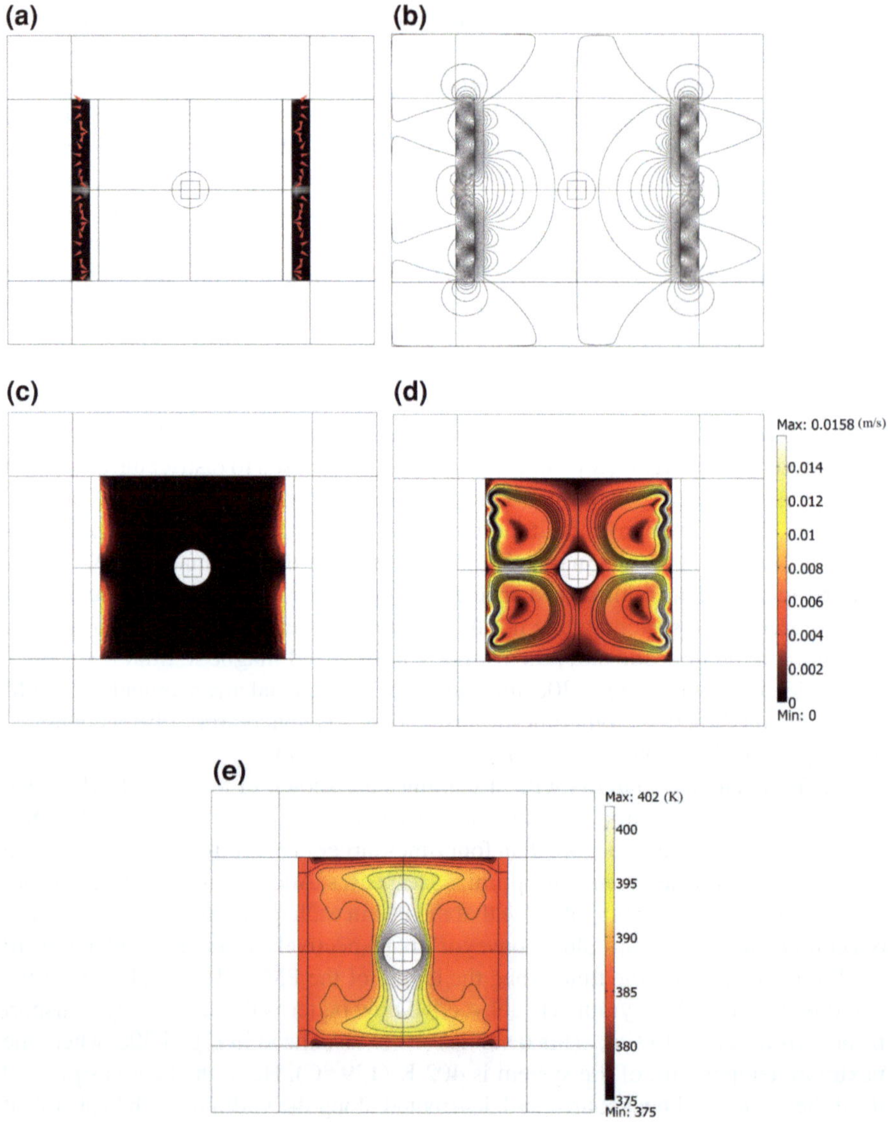

Fig. 5.30 Design optimization results for system from Fig 5.29: **a** PM distribution and magnetization direction (*note dark regions* = PM; *light regions* = air; *arrows* = magnetization direction); **b** magnetic vector potential field lines; **c** normalized magnitude of the magnetic fluid body force distribution (*note lighter regions* indicate larger magnitudes); **d** magnetic fluid velocity contours with streamlines; and **e** temperature contours. Reprinted from [60, Fig. 4], Copyright (2012), with permission from American Society of Mechanical Engineers (ASME)

Table 5.2 Four PM magnetization direction layouts (indicated by arrows) assigned for comparison with optimization results in Fig. 5.30

I: h sym x ant y	II: v par	III: v sym y ant x	IV: h ant y
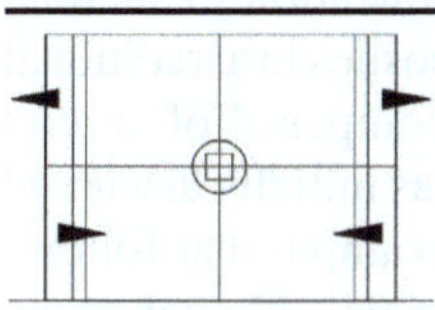	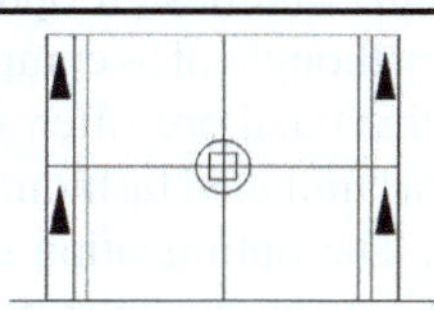	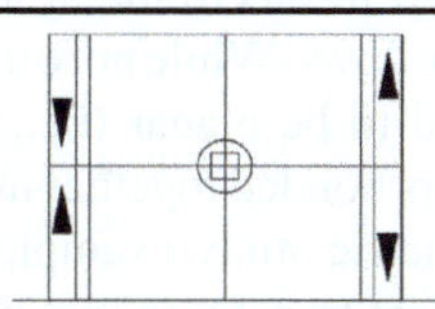	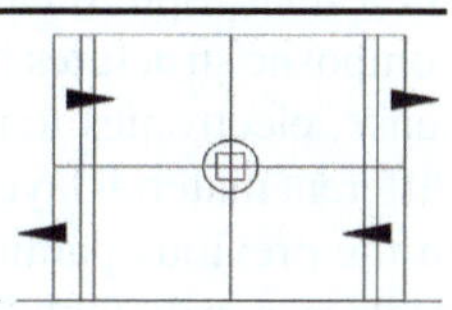

Note 'h' = horizontal; 'v' = vertical; 'par' = parallel; 'sym' = symmetric; 'ant' = anti-symmetric

Table 5.3 Fluid velocity contours plus magnetic vector potential lines for the cases from Table 5.2

I: h sym x ant y	II: v par	III: v sym y ant x	IV: h ant y
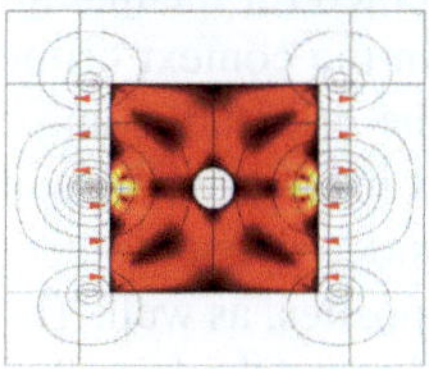	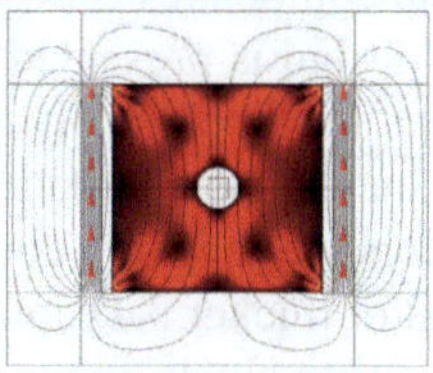	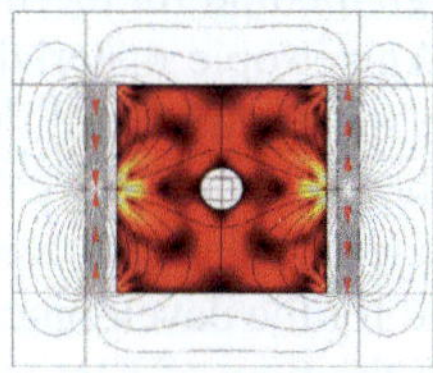	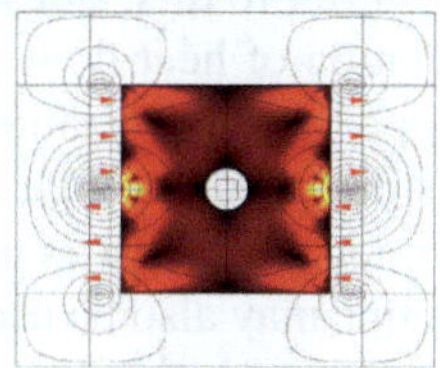
$T_{max} = 421.8$	$T_{max} = 422.7$	$T_{max} = 424.6$	$T_{max} = 432.4$
$v_{avg} = 7.20 \times 10^{-4}$	$v_{avg} = 5.44 \times 10^{-4}$	$v_{avg} = 6.85 \times 10^{-4}$	$v_{avg} = 5.40 \times 10^{-4}$
$v_{max} = 3.64 \times 10^{-3}$	$v_{max} = 1.57 \times 10^{-3}$	$v_{max} = 3.17 \times 10^{-3}$	$v_{max} = 3.24 \times 10^{-3}$

Note 'h' = horizontal; 'v' = vertical; 'par' = parallel; 'sym' = symmetric; 'ant' = anti-symmetric; 'rot' = rotating; 'K' = temperature units; 'm/s' = velocity units

the maximum temperature, T_{max}, average velocity magnitude of the fluid, v_{avg}, and maximum velocity magnitude of the fluid, v_{max}. The figures included in Table 5.3 show the corresponding fluid velocity contours with magnetic vector potential field lines and arrows denoting the PM magnetization directions.

The results in Table 5.3 indicate that an increased number of recirculation zones is generally not a good predictor of better heat transfer. For example, while the best performer is Case I, which has six recirculation zones, with a 421.8 K maximum temperature and 7.20×10^{-4} m/s average velocity, the second best performer (Case II), has four recirculation zones, with a 24 % lower average velocity and a maximum temperature that is only 0.9 K higher than Case I. Additionally, the average velocity for Case II is smaller in magnitude than the average velocity for Case III which also has six recirculation zones, yet Case II still results in a lower maximum temperature. The maximum velocity is even less of a predictor of good heat transfer performance. Furthermore, since the lowest maximum temperature achieved (i.e., Case I) is ~20 K higher than the optimized system shown in Fig. 5.30, intuition fails to provide an optimum PM layout, and in all likelihood, there exists no analytical method to solve this design problem.

5.1.5 Design Optimization of Thermal Composites

As a final topic related to electronic systems, the design optimization of thermal composite structures follows. While not often thought of as composites in a traditional sense, electronics tend to be planar (i.e., thin) and are often composed of several different material layers bonded together in a laminated fashion, as initially discussed in the previous parametric study example. The optimization example that follows deals primarily with heat transfer in two-phase composite materials, although many of the concepts may be extended to the multiphysics optimization of more complex electronic systems, where the associated methods are applicable to cases involving different physical processes and multifunctional composites. A brief discussion of these prospective research directions is provided at the end of this section.

Prior to this, a design method for thermal composites is presented for the manipulation of heat flow. Anisotropic material design is presented in the context of the topology optimization tools introduced thus far. Specifically, the orientation angle is designed for high thermal conductivity inclusions embedded in a lower thermal conductivity matrix material. Planar structures are generally considered, but the technique may also be used for the development of thicker 3-D composites, as well. The goal here is to control the flow of heat to either cloak, focus, or turn the heat flux in a given structure. Many potential applications exist for these so-called thermal metamaterials including the design of thermal circuits or multi-functional substrates for advanced electronics thermal management.

The governing equations for heat conduction presented in Chap. 3 once more apply here. However, since the mechanics of thermal composites have not been explicitly introduced, thus far, an expanded presentation of the governing equations is presented based on the authors' work in [30]. Then, the optimization model from [30] is introduced along with the associated structural designs obtained using different optimization objective functions.[3] Finally, related experimental results from [29] for ultra-thin composites are highlighted to validate the numerical results obtained using the optimization method, where the composites themselves represent typical substrate structures found in electronics such as printed circuit boards (PCBs).

Heat Conduction in Anisotropic Composites

Heat conduction in an isotropic solid is governed by Fourier's law [83]

$$Q = -\frac{\partial}{\partial x_i}\left(k\frac{\partial T}{\partial x_i}\right), \tag{5.43}$$

[3] With kind permission from [30].

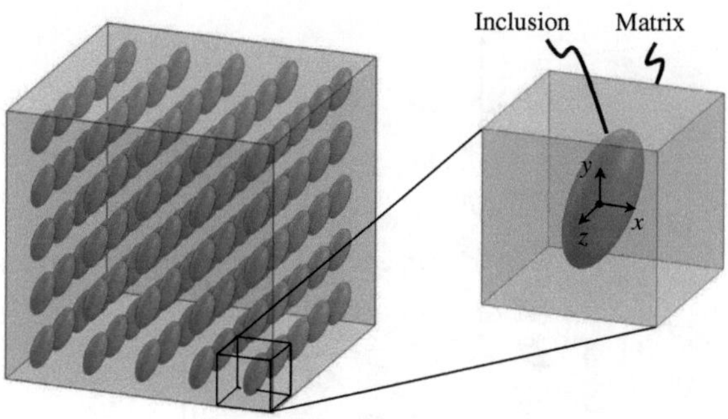

Fig. 5.31 Anisotropic thermal-composite material with *zoomed view* of unit cell showing a thermally conductive non-spherical inclusion embedded in a thermally conductive matrix. With kind permission from [30, Fig. 1], © Springer-Verlag Berlin Heidelberg 2013

where the thermal conductivity, k, is equal in all directions and the volumetric heat generated inside the domain, Q, flows in a direction perpendicular to the isothermal surface passing through the point in space under consideration [79].

In contrast, two-phase composite materials comprising nonspherical inclusions embedded in a surrounding matrix, such as the arrangement shown in Fig. 5.31, are expected to exhibit anisotropic material thermal conductivity, where heat conduction in the Cartesian coordinate system is governed by

$$Q = -\frac{\partial}{\partial x_i}\left(k_{ij}\frac{\partial T}{\partial x_j}\right). \tag{5.44}$$

For an anisotropic solid, the heat does not necessarily flow perpendicular to a given isothermal surface, and the thermal conductivity involves nine conductivity coefficients, k_{ij}, which are components of a symmetric second-order tensor,[4] **k**, [79] as follows:

$$k_{ij} = \begin{bmatrix} k_{11} & k_{12} & k_{13} \\ k_{21} & k_{22} & k_{23} \\ k_{31} & k_{32} & k_{33} \end{bmatrix}, \quad \text{where} \tag{5.45}$$

$$k_{ij} = k_{ji} \qquad i, j = 1, 2, 3. \tag{5.46}$$

In the case of a thin square plate structure with a temperature gradient, $-\tau_x$, along the x-axis shown in Fig. 5.32, the 3-D formulation above may be reduced to describe heat conduction in two dimensions (2-D). Here, the x-y plane dependence of the thermal conductivity on the inclusion angle, θ, may be approximated via the

[4] Note that **k** and **K** throughout this section, i.e., Sect. 5.1.5 refer, respectively, to the second-order thermal conductivity tensor and thermal conductivity matrix instead of local/global stiffness matrices.

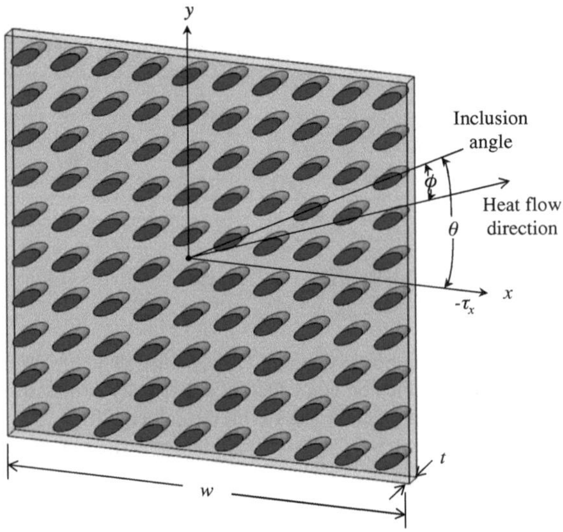

Fig. 5.32 A thin (i.e., $t << w$) square composite plate with anisotropic thermal conductivity comprising elliptic cylinder inclusions embedded in a surrounding matrix and subjected to a temperature gradient along the x-axis at an angle to the inclusion direction. With kind permission from [30, Fig. 2], © Springer-Verlag Berlin Heidelberg 2013

coordinate transformation [41]

$$k_{11} = k_1 \cos^2\theta + k_2 \sin^2\theta, \tag{5.47}$$

$$k_{22} = k_1 \sin^2\theta + k_2 \cos^2\theta, \tag{5.48}$$

$$k_{12} = (k_1 - k_2) \sin\theta \cos\theta, \tag{5.49}$$

where k_1 and k_2 are, respectively, the unit cell principal thermal conductivities parallel and normal to the inclusion direction.

In contrast to [22] where k_{12} was assumed to be zero, shear terms are now represented by Eq. (5.49). Furthermore, elliptic cylinder type inclusions were arbitrarily selected for the two-phase composite material microstructure; see Fig. 5.32. Thus, instead of using a simplified composite slab model [22], the unit cell principal k_1 and k_2 values may be determined for any shape and aspect ratio inclusion, per [42], via

$$\mathbf{K}_{uc} = \mathbf{K}_m + \frac{f_i\,(\mathbf{K}_i - \mathbf{K}_m)\,\mathbf{K}_m}{(1 - f_i)\,(\mathbf{K}_i - \mathbf{K}_m)\,\mathbf{S} + \mathbf{K}_m}. \tag{5.50}$$

In this expression, $\mathbf{K}_i$ is the thermal conductivity matrix for the isotropic inclusion, $\mathbf{K}_m$ is the thermal conductivity matrix of the surrounding isotropic media, f_i is the inclusion volume fraction, and $\mathbf{S}$ is a second-order tensor analogous to the Eshelby tensor [34, 42, 43]. For an elliptic cylinder type inclusion, expressions for the components of **S** in terms of the ellipse major, a_1, and minor, a_2, radii may be found in [42], and the inclusion aspect ratio may be defined as

$$\alpha_i = \frac{a_1}{a_2}. \tag{5.51}$$

This detailed microstructure formulation allows for the investigation of the effect of the inclusion aspect ratio on the guiding of heat flux within a composite. Also, note that while a two-phase composite with elliptic cylinder type inclusions was selected, [42] provides a microstructure formulation for three-phase systems as well as the **S** tensor components for a variety of other standard inclusion geometries.

As indicated in Fig. 5.32, the heat flow within the composite does not necessarily coincide with the direction of the temperature gradient, per [41], and the angle, φ, between the inclusion direction and temperature gradient may be found using

$$\tan\varphi = (k_{22}/k_{11})\tan\theta. \tag{5.52}$$

This expression may be used to accurately compute the heat flow inside the composite material.

Formulation of the Optimization Problem

The formulation of the material thermal conductivity design optimization problem follows the detailed numerical procedure provided in [22]. A standard finite element solver [15] is utilized for the steady-state conduction heat transfer problem, and a gradient-based Method of Moving Asymptotes optimizer [90] is used to design the material microstructure by determining the optimal inclusion angle, θ, that minimizes a particular objective function.

To find the optimal inclusion angle distribution for the anisotropic material thermal conductivity, the optimization problem is formulated as

$$\begin{array}{rl} \text{Find} & \gamma \\ \text{Minimize} & \mathsf{F}_o \\ \text{Subject to} & \text{Eq. (5.44)} \\ & -1 \le \gamma \le 1 \end{array}$$

Here, F_o is a single or multiterm objective function, and the material design variable, γ, is constrained on the interval $[-1, 1]$. Since the scalar function γ can take any point-wise value, the raw function may produce oscillating designs and a Helmholtz PDE-based filter is implemented in the optimization routine, per [51]. This filter acts to produce a smooth scalar function, $\tilde{\psi}$, that is then transformed into θ in Eqs. (5.47)–(5.49) to cover the range of inclusion design angles covering $-180°$ to $+180°$; i.e., $\theta \cong \tilde{\psi}\pi$. Thus, the entire optimization problem is formulated as follows:

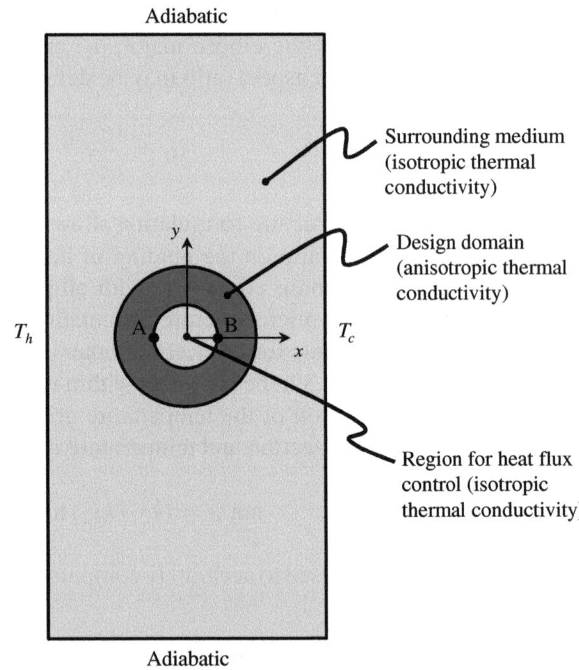

Fig. 5.33 Benchmark problem from [70] considered in the formulation of the optimization objective function for heat flux shielding, focusing, and reversal. Heat flux in the *center* domain inside the ring is manipulated by designing the ring anisotropic material thermal conductivity. With kind permission from [30, Fig. 3], © Springer-Verlag Berlin Heidelberg 2013

$$
\begin{aligned}
\text{Find} \quad & \gamma \\
\text{Minimize} \quad & \mathsf{F}_o \\
\text{Subject to} \quad & \text{Eq. (5.44)} \\
& \text{Eqs. (5.47)} - (5.49) \\
& \theta = \tilde{\psi}\pi \\
& -R_f^2\nabla^2\tilde{\psi} + \tilde{\psi} = \gamma \\
& -1 \leq \gamma \leq 1
\end{aligned}
$$

where R_f is the filter radius for Helmholtz filtering [51].

To facilitate the introduction of the different optimization objective functions for heat flux shielding, focusing, and reversal, a schematic of a benchmark design optimization example is provided in Fig. 5.33. Here, an anisotropic composite ring is placed in a surrounding isotropic thermally conductive medium as proposed in [70], where the design domain consists of the ring itself and a left-to-right temperature gradient is created by enforcing a uniform fixed hot side temperature, T_h, on the left vertical boundary plus a uniform fixed cold side temperature, T_c, on the right vertical boundary of the surrounding medium. The top and bottom boundaries of the surrounding domain are considered adiabatic.

Objective Functions for Heat Flux Control

In this section, a general energy form of the objective function for heat flux control is proposed followed by related objective functions for heat flux shielding and focusing. For heat flux reversal, a multi-term objective function is developed that utilizes the directionality of the thermal gradient.

General Energy Formulation

Following [11, 22], a general energy formulation of the objective function for the minimization of the thermal compliance of a 2-D domain comprising a material having anisotropic thermal conductivity may be given as

$$\mathsf{F}_{o\_g} = k_{11}\frac{\partial^2 T}{\partial x^2} + k_{22}\frac{\partial^2 T}{\partial y^2} + 2k_{12}\frac{\partial T}{\partial x}\frac{\partial T}{\partial y}, \tag{5.53}$$

where the terms k_{11}, k_{22}, and k_{12} in this expression are defined in Eqs. (5.47)–(5.49). In contrast to [22] where an orthotropic material was assumed with $k_{12} = 0$, shear terms are now represented by the third term on the right-hand side (RHS) in Eq. (5.53).

Heat Flux Shielding

Heat flux shielding may be realized by minimizing the magnitude of the temperature gradients and heat flow within a selected region of a system. It can be achieved by minimizing thermal compliance, F_{o_g}, in Eq. (5.53). Thus, the objective function for shielding, F_{o_s}, may be adopted as

$$\mathsf{F}_{o\_s} = \mathsf{F}_{o\_g}. \tag{5.54}$$

Note that in the case of the center domain with isotropic thermal conductivity, k_o, shown in Fig. 5.33, $k_{11} = k_{22} = k_o$ and $k_{12} = 0$ resulting in the simplified objective function

$$\mathsf{F}_{o\_s\_i} = k_o\left(\frac{\partial^2 T}{\partial x^2} + \frac{\partial^2 T}{\partial y^2}\right). \tag{5.55}$$

Heat Flux Focusing

Heat flux focusing involves the maximization of the thermal compliance or temperature gradient (and heat flow) within a select region of the system. The objective function for heat flux focusing may be expressed by placing a minus sign in front of Eq. (5.53)

$$\mathsf{F}_{o\_f} = -\mathsf{F}_{o\_g}, \tag{5.56}$$

and in case of the center domain with isotropic thermal conductivity in Fig. 5.33, a simplified objective function is likewise obtained

$$\mathsf{F}_{o\_f\_i} = -\mathsf{F}_{o\_s\_i}. \tag{5.57}$$

Heat Flux Reversal

In the case of heat flux reversal, the direction and magnitude of the temperature gradient and heat flow are important. Here, Fourier's law for an anisotropic composite may be written in the following form:

$$\mathbf{q} = -\mathbf{k}\nabla T, \tag{5.58}$$

where $\mathbf{q}$ is the heat flux vector at a spacial position. Expanding the RHS of Eq. (5.58) gives

$$\mathbf{q} = -\begin{bmatrix} k_{11}\frac{\partial T}{\partial x} + k_{12}\frac{\partial T}{\partial y} \\ k_{21}\frac{\partial T}{\partial x} + k_{22}\frac{\partial T}{\partial y} \end{bmatrix}. \tag{5.59}$$

Assuming the heat flux vector, $\mathbf{q}$, is aligned with a unit vector, $\hat{\mathbf{e}}$, that lies along a target angle direction in the global coordinate system; Eq. (5.59) suggests that the direction of heat flow through a given spatial position may be controlled by defining an objective function that realigns the heat flux vector with a target unit vector, $-\hat{\mathbf{e}}$, pointed in the opposite direction. Thus, the following multi-term objective function is proposed:

$$\mathsf{F}_{o\_r} = w_1 T_1 + w_2 T_2, \tag{5.60}$$

where w_1 and w_2 are assumed weighting values and T_1 and T_2 are given by

$$T_1 = \hat{\mathbf{e}} \cdot \mathbf{q}, \quad \text{and} \tag{5.61}$$

$$T_2 = \left|\hat{\mathbf{e}} \times \mathbf{q}\right|^2. \tag{5.62}$$

Physically, heat flux directivity is achieved by taking the scalar product of $\hat{\mathbf{e}}$ and $\mathbf{q}$ via Eq. (5.61). By minimizing T_1, the component of $\mathbf{q}$ that is anti-parallel to $\hat{\mathbf{e}}$ is maximized, However, this criteria does not strictly make $\mathbf{q}$ anti-parallel to $\hat{\mathbf{e}}$ since it does not minimize the normal component of $\mathbf{q}$ directly. In order to fully align the angle of $\mathbf{q}$ to $\hat{\mathbf{e}}$, the normal component should be minimized. By including the norm of the cross product via Eq. (5.62), the angular error between $\hat{\mathbf{e}}$ and $\mathbf{q}$ is further reduced (i.e., $\left|\left\|\hat{\mathbf{e}}\right\| \|\mathbf{q}\| \sin(\vartheta)\, \hat{\mathbf{n}}\right| \to 0$ as $\vartheta \to 0$ or π).

Numerical Studies

In this section, the validity of the objective functions from the previous section for heat flux shielding, focusing, and reversal is examined via numerical experiments involving the benchmark problem from Fig. 5.33. The goal is to control the flow of heat within the isotropic center domain of Fig. 5.33 by designing the structural layout of the anisotropic composite ring. Comparisons are made with the results reported in [70].

Uniform hot and cold side temperature boundary conditions of 314 K and 273 K, respectively, were applied to the structure shown in Fig. 5.33. As described in [70], for the composite ring to blend into the background and not disturb the external

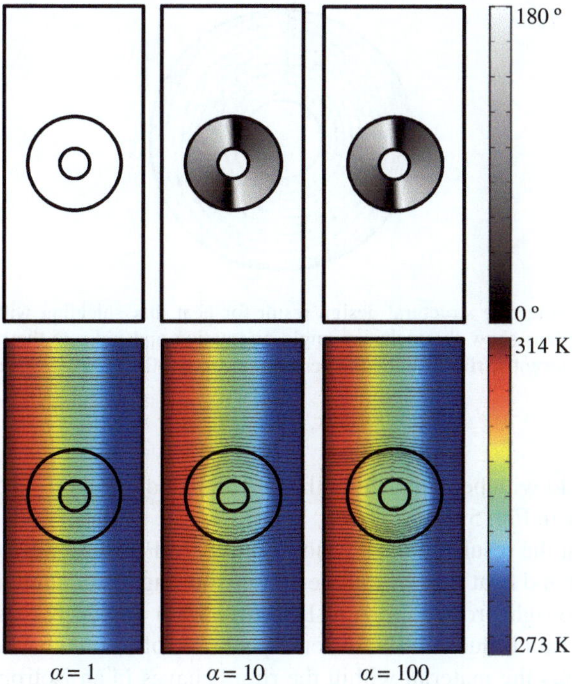

Fig. 5.34 Composite ring heat flux shielding design results. The results in *left*, *middle*, and *right* columns correspond to inclusion aspect ratios of $\alpha_i = 1$, 10, and 100, respectively. The images in the *top row* show the absolute value of the composite structure design variable field. The images in the *bottom row* show the temperature field with heat flux streamlines. With kind permission from [30, Fig. 4], © Springer-Verlag Berlin Heidelberg 2013

temperature field, the thermal resistance of the surrounding medium (including center region) should be close to the reduced average of the materials comprising the two-phase ring, i.e., $k_i k_m \sim k_o^2$. Thus, similar values of 2.6 W/(m K), 0.13 W/(m K), and 0.56 W/(m K) were chosen for the thermal conductivity of the inclusion, matrix, and surrounding media, respectively. An inclusion volume fraction of 0.5 with initial vertical orientation was assumed, and various inclusion aspect ratios were examined in the numerical examples that follow. For all of the numerical results, a filter radius, $R_f = 0.002$, was used, and the optimization objective function was defined exclusively on the center domain enclosed by the composite design domain.

Design Results: *Heat Flux Shielding*

The images in the left, center, and right columns in Fig. 5.34 show the heat flux shielding design results for the composite ring using inclusion aspect ratios of $\alpha_i = 1$, 10, and 100, respectively. The top row in Fig. 5.34 shows the absolute value of the structural design result, where dark colored regions indicate an inclusion angle of zero degrees and light colored regions indicate an inclusion angle of 180°. The

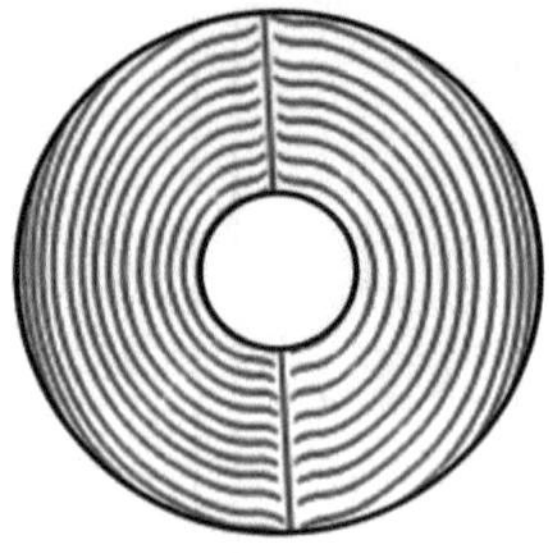

Fig. 5.35 Composite ring structural design result for heat flux shielding with $\alpha_i = 100$. The computed streamlines show the inclusion angle orientation and indicate that the inclusions are aligned to form concentric rings. With kind permission from [30, Fig. 6], © Springer-Verlag Berlin Heidelberg 2013

temperature field with heat flux streamlines, computed using Eq. (5.52), is shown in the bottom row in Fig. 5.34.

Observe that the results in the leftmost column of Fig. 5.34 were obtained after a single iteration and heat flux within the composite ring is not altered; a temperature gradient (left to right from point A to B in Fig. 5.33) of 7.24 K exists in the center domain. This result is due to an inclusion aspect ratio of $\alpha_i = 1$ that defines a circular particle, and thus the material within the ring behaves in an isotropic fashion with heat flux unmodified within the ring, as expected. As the particle aspect ratio is increased to 10 and then 100, the solution shows increased heat flux shielding (as evidenced by the streamlines in Fig. 5.34) with reduced temperature gradients in the center domain of 3.33 K and 2.07 K, respectively.

In the limit, as $\alpha_i \rightarrow \infty$, the k_1 and k_2 principal thermal conductivity values defined by the composite microstructure model, Eq. (5.50), approach values obtained using a layered slab approach; refer to [22]. Thus, for higher inclusion aspect ratios the solution for the composite ring quickly converges toward a structure comprising approximately concentric laminations, as shown in Fig. 5.35. This structure matches the composite ring design originally proposed in [70], where the temperature field outside of the ring is unaffected and all of the heat flux manipulation occurs inside the composite.

Design Results: *Heat Flux Focusing*

Several inclusion aspect ratios were examined for the case of heat flux focusing, where the optimal structure was found using Eq. (5.56) after approximately 50 iterations. The temperature gradient (left to right from point A to B in Fig. 5.33) in the center domain for $\alpha_i = 10$ and 100 was 12.4 K, and 14.6 K, respectively, with over 100 % increase in the heat flux for the higher aspect ratio. For reference, the design results for $\alpha_i = 100$ are shown in Fig. 5.36, where again the temperature contours outside of the ring are unaffected and all of the bending occurs within the composite ring. Higher aspect ratio inclusions lead to a greater amount of heat flux focusing, and

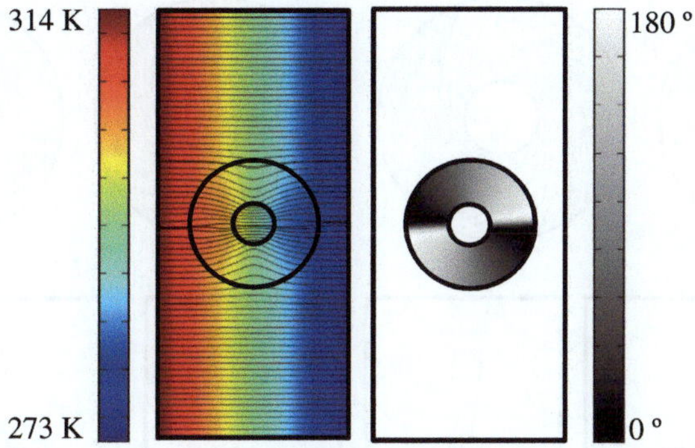

Fig. 5.36 Composite ring heat flux focusing design results with inclusion aspect ratio of $\alpha_i = 100$. The image on the *left* shows the temperature field with heat flux streamlines. The image on the *right* shows the absolute value of the composite structure design variable field. With kind permission from [30, Fig. 7], © Springer-Verlag Berlin Heidelberg 2013

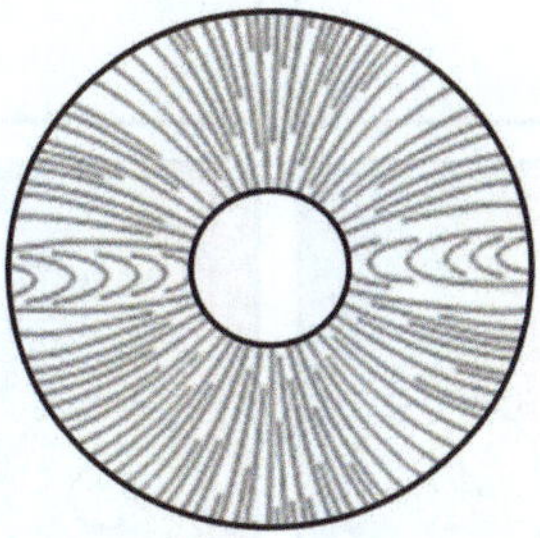

Fig. 5.37 Composite ring structural design result for heat flux focusing with $\alpha_i = 100$. The computed streamlines show the inclusion angle orientation and indicate that the inclusions are aligned to form radial spokes. With kind permission from [30, Fig. 8], © Springer-Verlag Berlin Heidelberg 2013

the resultant optimal structure in Fig. 5.37 resembles a composite ring composed of radially extending laminated spokes, which is again similar to the assumed structure for heat flux concentration in [70].

Design Results: Heat Flux Reversal

The structural design results for heat flux reversal are shown in Fig. 5.38. For the case of heat flux reversal, it was found that an order-of-magnitude larger inclusion thermal conductivity value, 50 W/(m K) instead of 2.6 W/(m K), was required to trigger a twisting effect in the structural design result. Also, higher inclusion aspect ratios were found to produce a greater heat flux reversal effect, and specifying the weighting values w_1 and w_2 along with the target angle (i.e., $\hat{\mathbf{e}}$-vector) in Eqs. (5.60)–(5.62) allowed for specific modification of the twist angle.

Each set of results was obtained within 100 numerical iterations using $\alpha_i = 10^4$. The images in the columns of Fig. 5.38 (from left to right) are for different weighting

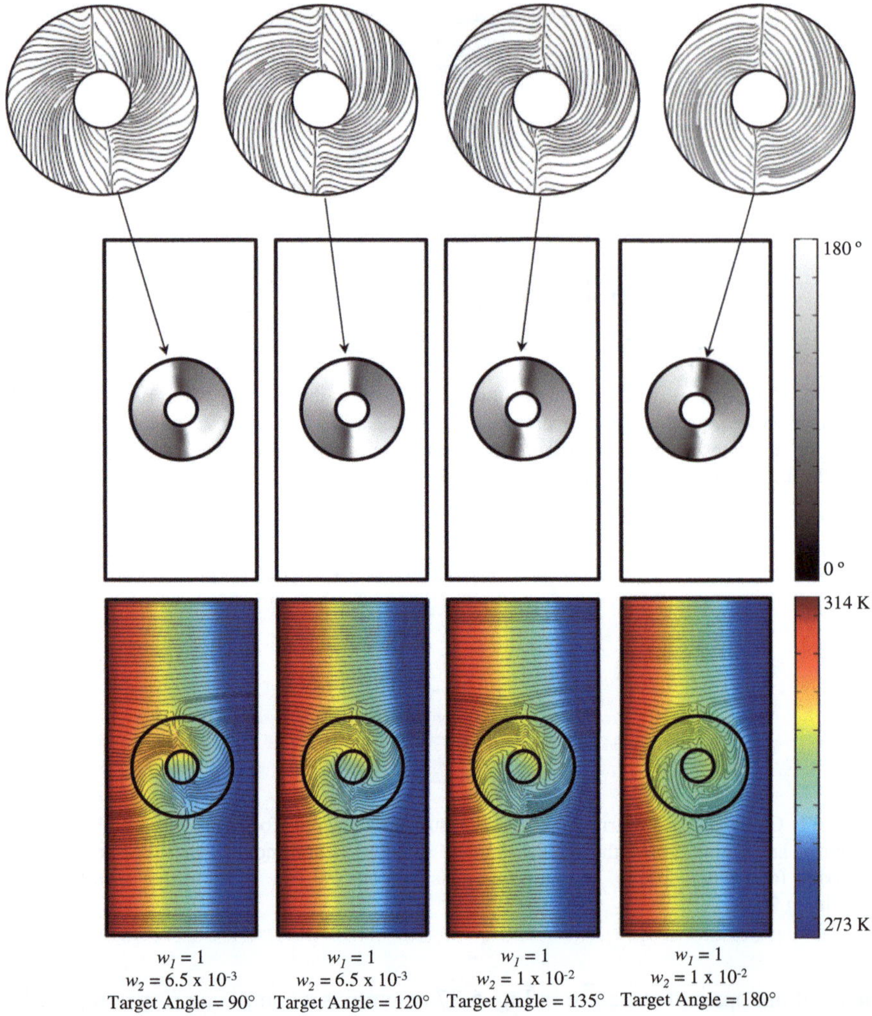

Fig. 5.38 Composite ring heat flux reversal design results with inclusion aspect ratio of $\alpha_i = 10^4$. The results in columns correspond to different w_1, w_2, and target angle (i.e., $\hat{\mathbf{e}}$-vector) parameters in Eqs. (5.60)–(5.62). The computed streamlines in the images in the *top row* show the inclusion angle orientation and indicate that the inclusions are aligned to form approximate spiral structures. The images in the *middle row* show the absolute value of the composite structure design variable field. The images in the *bottom row* show the temperature field with heat flux streamlines. With kind permission from [30, Fig. 9], © Springer-Verlag Berlin Heidelberg 2013

values and target angle combinations that produce a 90°, 120°, 135°, and 180° turn in the heat flow. Complementary views of the composite structural design results are shown in the top and middle rows of Fig. 5.38, while the temperature field with heat flux streamlines is shown in the bottom row.

Inspection of the lower left temperature field image in Fig. 5.38 reveals that prioritizing heat flow in the vertical direction ($w_1 = 1; w_2 = 6.5 \times 10^{-3}; \hat{\mathbf{e}} = [0 \quad 1]$) produces an exact 90° turn in the heat flux. The corresponding structural design results (left middle and top images) show that a spiral shape laminated structure is formed to guide the heat flux across the center domain. However, the spiral is not fully wrapped, and the upper left and lower right corners of the ring are designed such that the inclusions follow a discontinuous path that guides the heat flux vertically inward.

Moving to the second column from the left in Fig. 5.38, by specifying $w_1 = 1$, $w_2 = 6.5 \times 10^{-3}$, and $\hat{\mathbf{e}} = [0.5 \quad 0.866]$, the composite ring is designed to produce a 120° turn in the direction of the heat flux. Here, the spiral shape structure is less discontinuous in the upper left and lower right regions of the ring, which creates a smoother heat flow path around the annulus. This smoothing of the spiral continues as seen in the third column from the left in Fig. 5.38, where the parameters $w_1 = 1$, $w_2 = 1 \times 10^{-2}$, and $\hat{\mathbf{e}} = [0.707 \quad 0.707]$) generate a 135° turn in the direction of the heat flux.

Finally, a complete 180° heat flux reversal across the center domain is obtained by setting $w_1 = 1$, $w_2 = 1 \times 10^{-2}$, and $\hat{\mathbf{e}} = [1 \quad 0]$) in Eqs. (5.60)–(5.61); refer to the lower right temperature field image in Fig. 5.38 where a temperature gradient (left to right from point A to B in Fig. 5.33) of −4.3 K exists. Observe in the corresponding structural design results (right middle and top images in Fig. 5.38) that the arrangement of the inclusions is nearly continuous and approximates a logarithmic spiral. Despite some bending of the temperature field outside of the composite ring due to the larger assumed inclusion thermal conductivity value, this structural design result matches the layout of the 'fan-like' composite from [70] which was obtained via an assumed coordinate transformation approach.

5.1.5.1 Extensions to Prototype Design

While the fundamental experimental results in [70] provide a satisfactory level of validation for the presented optimization method for heat flux shielding (i.e., cloaking), focusing, and reversal, the demonstration of such devices in ultra-thin composite structural embodiments is a critical step to implementation in practical electronics systems. The details of such representative structures may be found in [29], and are summarized in this section. Specifically, two-material composite structures fabricated using standard PCB manufacturing techniques are described. These example composites were designed for heat flux cloaking, focusing, and reversal by etching high thermal conductivity copper trace patterns in an engineered fashion on low thermal conductivity PCB substrates. A schematic (including dimensions) of such a PCB structure for heat flux cloaking is provided in Fig. 5.39a, where the left and right ends of the structure are designed as thermal bus bars for the application and removal of heat, respectively, to the structure. The center region of the structure consists of concentric copper ring traces, as suggested by the design results shown in Fig. 5.35, while the surrounding region comprises a copper mesh designed to cloak the cen-

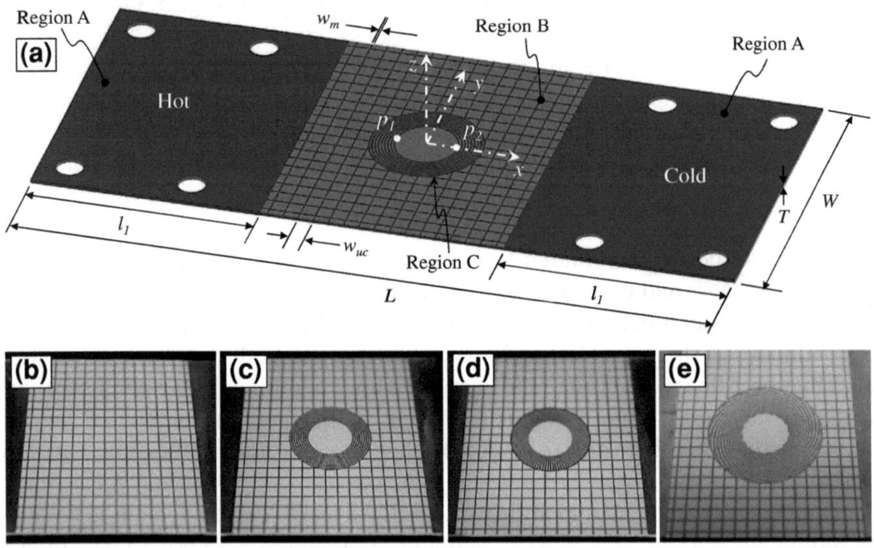

Fig. 5.39 **a** A representative schematic of an ultrathin anisotropic composite device for heat flux cloaking. *Dimensions* device length, L = 115 mm; device width, W = 50 mm; device thickness, T =578 μm; mesh width, w_m =200 μm; unit cell width, w_{uc} = 2.5 mm; and thermal bus bar length, l_1 = 37.5 mm. *Zoomed view* image of the *center* region of each fabricated composite: **b** baseline device without *center* heat flow control region, **c** heat flux cloaking device (ring OD ~1.85 cm, ID ~1 cm), **d** heat flux focusing device (ring OD ~1.85 cm, ID ~1 cm), and **e** heat flux reversal device (ring OD 2.5 cm, ID ~1 cm). Reprinted, with permission, from [29, Fig. 1], Copyright (2013), AIP Publishing LLC

ter heat flow manipulation region; refer to [29] for details. In Fig 5.39b, a baseline fabricated composite structure without a center heat flow control region is shown, while Figs. 5.39c, d, and e, show PCB structures for heat flux cloaking, focusing, and reversal, respectively. The focusing and cloaking composites, respectively, utilize radial spoke traces and spiral traces in the center heat flux manipulation region consistent with the design results shown in Figs. 5.37 and 5.38.

As explained in [29], a fixed temperature gradient of 35 K was applied across the center region of each device using a copper block plus cartridge heater mounted to the hot side thermal bus bar and a thermoelectric cooler connected to the cold side thermal bus bar. The temperature field of the center region of each device was then imaged using a calibrated infrared camera. Additional heat transfer simulations were performed to verify the response of each PCB composite considering both conduction and convection effects. The numerical and experimental results are shown in Fig. 5.40 with computational results shown on the top and experimental results shown on the bottom; refer to [29] for further details.

Observe in Fig. 5.40 that the measured response of each thermal composite is consistent with the heat flux cloaking, focusing, and reversal effects anticipated in the prior optimization study. These results indicate that such anisotropic heat

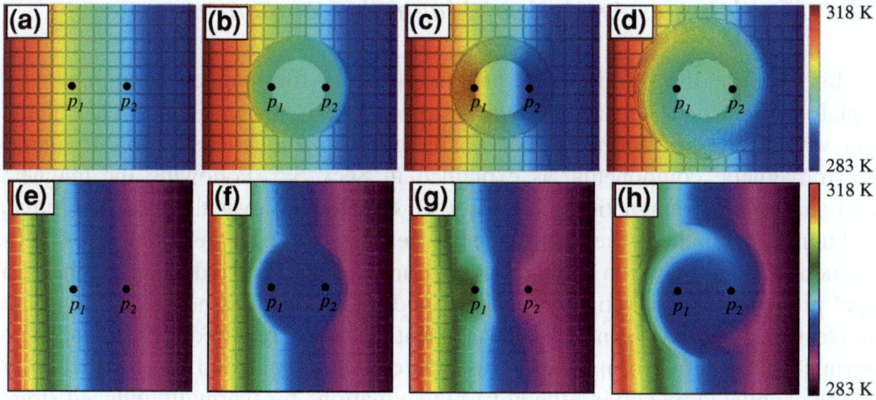

Fig. 5.40 Thermal contours of each device center region: **a** baseline—numerical, **b** heat flux cloaking—numerical, **c** heat flux focusing—numerical, **d** heat flux reversal—numerical, **e** baseline—experimental, **f** heat flux cloaking—experimental, **g** heat flux focusing—experimental, and **h** heat flux reversal—experimental. *Note* points p_1 and p_2 indicate locations for temperature gradient measurement; refer to Fig. 5.39. Reprinted, with permission from [29, Fig. 3], Copyright (2013), AIP Publishing LLC

conduction models and associated optimization tools may be used with good accuracy in the development of advanced PCB thermal composites for electronics thermal management applications. Logical ongoing extensions of this work include the co-design of multilayer PCBs for electrothermal functionality as well as laminated composites for a variety of electromechanical applications in consumer electronics, automotive, and aerospace fields.

5.2 Magnetic Component Analysis and Design

Magnetic components (inductors and transformers) are essential in the construction of electric circuits. They control the circuit current and voltage during the electric and magnetic energy conversion process. Multiphysics analysis might enable us to predict the electrical, magnetic, structural and thermal performance of these components. The main performance metrics for magnetic components are inductance and heat loss. The inductance indicates the amount of magnetic energy generated by electrical currents, which will in turn determine the induced voltage in an electrical circuit. Heat losses also occur during the aforementioned energy conversion process, and these losses determine the efficiency and thermal performance of magnetic components. The design optimization of a magnetic component enables component size reduction while maintaining electric and magnetic performance or improving the component thermal performance. Accordingly, the multiphysics analysis procedure for magnetic components is presented in this section followed by a 2-D analysis example for an inductor. Additionally, a design optimization example for an inductor is presented.

5.2.1 Multiphysics Analysis of Magnetic Components

In this section, the multiphysics analysis process for magnetic components is explained. The first step in the analysis procedure is the magnetic field calculation, which can be performed by solving the governing equations for low-frequency electromagnetic components, as explained in Sect. 3.5. Following the magnetic field analysis, the magnetic flux density, B, distribution over the analysis domain is obtained. From there, the inductance and heat losses can be predicted based on the magnetic field distribution and the electric current input. The inductance is the principal property for magnetic components, and inductance is a necessary parameter for electrical circuit designers. Likewise, heat losses are required for the analysis of thermal-structural performance. The associated temperature and stress distributions can be calculated by solving the governing equations for thermally induced stress, as discussed in Sect. 3.3. The detailed process for the calculation of inductance and heat losses is presented below.

Inductance Calculation

Once the magnetic field analysis is complete, the inductance, L, may be calculated from the distribution of the magnetic flux density, B, using one of two methods, as explained below. The first method is derived from the definition of magnetic flux linkage, while the second method is derived from a magnetic energy formulation.

Method 1—Magnetic Flux Linkage The first method utilizes magnetic flux linkage, λ_{link}, for the calculation of the inductance, L. The magnetic flux, Φ_B, is defined as $\int Bda$, where B is the magnetic flux density and a is the area of a winding loop. Then, the magnetic flux linkage may be expressed in terms of the magnetic field as

$$\lambda_{\text{link}} = N_c \Phi_B = N_c \int Bda, \tag{5.63}$$

where N_c is the number of coil winding turns. In terms of an electrical circuit, the magnetic flux linkage is again defined as

$$\lambda_{\text{link}} = Li, \tag{5.64}$$

where i is the coil electrical current. By equating Eqs. (5.63) and (5.64), the inductance is then derived as

$$L = \frac{N_c}{i} \int Bda. \tag{5.65}$$

It should be noted that the magnetic flux density, B, inside the winding area is utilized in the calculation of Eq. (5.65).

Method 2—Magnetic Energy In the second approach, the inductance is calculated using a magnetic energy formulation. The magnetic energy, W_{mag}, in a linear material may be assumed to be

$$W_{\text{mag}} = \frac{1}{2\mu} \int B^2 dv, \tag{5.66}$$

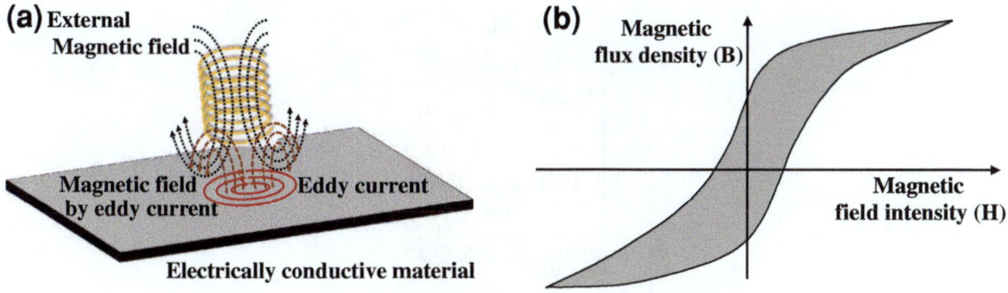

Fig. 5.41 The two-loss phenomena of ferromagnetic material in an inductor core: **a** eddy current losses are due to induced eddy currents in the conductive material which generates Joule heating losses; **b** hysteresis losses are represented by the total shaded area encompassed within the **B-H** curve

where μ is the magnetic permeability and v is the volume of the analysis domain. The magnetic energy of the electric circuit is defined as

$$W_{\text{mag}} = \frac{1}{2} L i^2. \tag{5.67}$$

By equating the two energy formulations, Eqs. (5.66) and (5.67), the inductance is derived as

$$L = \frac{1}{\mu i^2} \int B^2 dv \tag{5.68}$$

Here, the magnetic flux density, B, in the entire analysis domain is used for the calculation.

Heat Loss Calculation

The heat losses in magnetic components can be classified into coil losses (i.e., load losses) and core losses (i.e., no-load losses). Coil losses are related to Joule heating, which results from the flow of the electric current through the coil, that is, the electrically conductive material. The coil power loss, $\mathscr{P}_{\text{coil}}$, in Watts can be represented as

$$\mathscr{P}_{\text{coil}} = R i^2, \tag{5.69}$$

where R is the electric resistance of the coil. As the operating frequency of a magnetic component increases, the electrical resistance increases due to skin and proximity effects, which occur due to the magnetic field passing through the coil winding area.

As shown in Fig. 5.41, core loss is related to two separate effects that arise when dealing with any ferromagnetic core material: (1) eddy current effects, and (2) hysteresis effects. Eddy current effects develop when electric currents are induced within an electrically conductive material due to a change in the magnetic field. Associated eddy current power loss, $\mathscr{P}_{\text{eddy}}$, is related to Joule heating as follows:

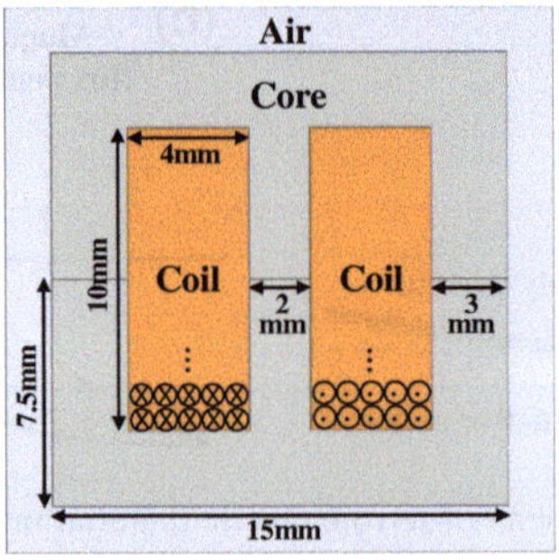

Fig. 5.42 Magnetic component analysis example: a 2-D inductor model comprising two E-shaped core components and coils

$$\mathscr{P}_{\text{eddy}} = \int \left(\sum_p k_e f_p^2 B_p^2 \right) dv, \tag{5.70}$$

where k_e is a coefficient related to the electrical conductivity of the core material, f_p is the p-th sinusoidal operating frequency, B_p is the p-th sinusoidal magnetic flux density, and v is the domain volume of core material.

Hysteresis effects for ferromagnetic materials result in heat loss and are related to the path dependency of the prior state of the magnetic flux density, B, versus the magnetic field intensity, H, curves. Hysteresis loss can be expressed as

$$\mathscr{P}_{\text{hysteresis}} = \int \left(\sum_p k_h f_p^2 B_p^n \right) dv, \tag{5.71}$$

where k_h is a coefficient determined using the hysteresis curve of the core material, n is also a material coefficient usually between 1.5 and 2.5, and v is the domain volume of the core material. The total core loss is then calculated as the sum of eddy current losses, Eq. (5.70), and hysteresis losses, Eq. (5.71), which can be calculated using the amplitude of p-th sinusoidal variation of magnetic flux density, B_p, inside the core material. Finally, the summation of the coil loss and total core losses produces the total loss, $\mathscr{P}_{\text{loss\_total}}$, of a given magnetic component

$$\mathscr{P}_{\text{loss\_total}} = \mathscr{P}_{\text{coil}} + \mathscr{P}_{\text{eddy}} + \mathscr{P}_{\text{hysteresis}}. \tag{5.72}$$

5.2.2 Analysis Example: 2-D Inductor Model

The 2-D multiphysics analysis of an E-shaped core inductor is presented as an analysis example for magnetic components. The assumed planar geometry of the E-shaped core inductor is shown in Fig. 5.42. The size of the core is arbitrarily selected, and

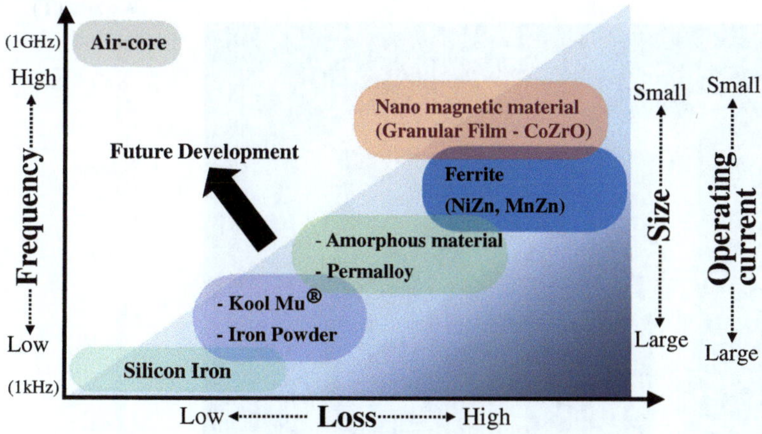

Fig. 5.43 Frequency versus core loss for existing inductor core materials

Table 5.4 Material physical parameters for inductor analysis example

Components (Material)	Relative magnetic permeability μ_r	Electric conductivity ς (S/m)	Thermal conductivity k (W/m K)
Coil (copper)	1	5.988×10^7	400
Core (steel)	900	1×10^4	44.5
air	1	0	–

the air-gap between two identically sized E-shaped core pieces is set to zero. The thickness of the inductor core is defined as 10 mm. The coil is wound using 60 turns of AWG 15 wire (diameter = 1.450 mm). The coil current is set as 10 A, and the frequency of the current is set to 20 kHz.

Figure 5.43 shows an assessment of existing core materials for inductor design. The inductor core loss is basically proportional to the device operating frequency since eddy current and hysteresis losses are a function of the operating frequency; refer to Eqs. (5.70) and (5.71). As the frequency increases, the inductor size can be miniaturized because the reactance is proportional to the frequency. However, the operating current and power is then limited due to increased core loss. Thus, an air-core may be used for an inductor that operates at a very high frequency (e.g., GHz regime).

Table 5.4 shows the assumed material physical parameters of the inductor components, which are needed for the magnetic and thermal analysis in this example. A magnetic analysis with harmonic excitation is performed by solving the governing equations explained in Sect. 3.5. In the analysis, an air-region is included since the permeability of the air and core material is of the same order, and an insulating condition is applied at the air boundary. From the analysis, the magnetic field distribution is obtained. Then, the inductance and heat losses are calculated using the method explained in Sect. 5.2.1. The thermal analysis is performed by solving the governing

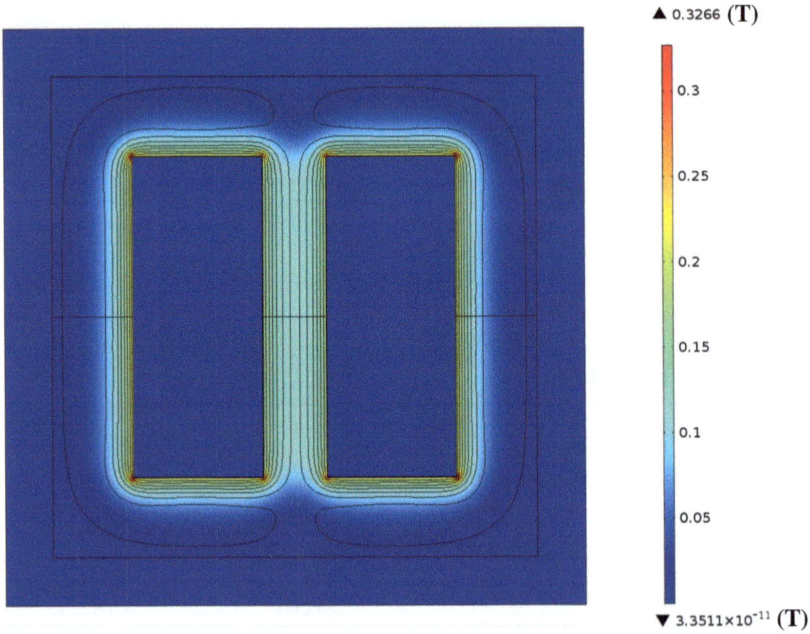

Fig. 5.44 Two-dimensional magnetic field analysis result for the E-shaped core inductor of Example 1. The *color contours* show the strength of the magnetic flux density, while the *lines* indicate magnetic equipotential lines

heat diffusion equation introduced in Sect. 3.2. For simplicity in the thermal analysis considered here, the air-region is excluded and a fixed temperature condition is applied at the boundary of the core material (assumed here to be 293 K). However, a convection heat transfer boundary condition could also be applied to the edge of the core material, or the model could be further coupled to a third physical process and include natural convection via a conjugate heat transfer analysis that includes the surrounding air domain.

The numerical model was discretized using 6,762 second-order triangular elements, and the coupled magnetic and thermal analysis required approximately 5 s to solve on a quad-core workstation with a 3.4 GHz processor and 16 GB of RAM. The magnetic field distribution for the E-shaped core inductor is presented in Fig. 5.44. Due to the eddy currents inside the core material, the magnetic field tends to be concentrated near the surface of the core material. The maximum magnetic flux density, 0.327 T, appears around the corner between the core and coil area. The inductance is calculated using both methods explained above. The inductance is calculated as 0.331 μH using the first method based on the flux linkage formulation, Eq. (5.65), and the inductance is 0.365 μH as determined using the second method based on the magnetic energy formulation, Eq. (5.68). The two inductance calculation methods give similar results. It should be noted again that the disadvantage of the first method is that the inductance calculation result can vary depending on the domain defined for

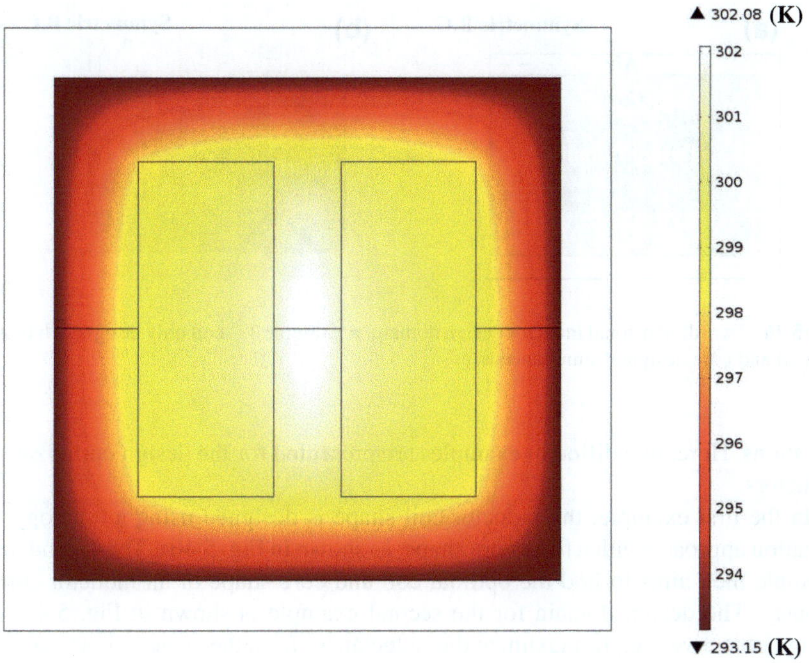

Fig. 5.45 Two-dimensional thermal analysis result for the E-shaped core inductor of Example 1. The *color contours* show the temperature distribution

the flux linkage calculation. Thus, the second method might be preferred in that the finite element analysis is based on an energy formulation, and the magnetic energy for the entire domain is used for the calculation (i.e., there is no need to decide which domain to use).

From the magnetic field distribution, the eddy current and hysteresis losses are, respectively, calculated using Eqs. (5.70) and (5.71). In this example, the material-dependent parameters k_e and k_h, and n in Eqs. (5.70) and (5.71) are, respectively, set to 3.5, 1.5 and 1.8. The coil loss is simply calculated via Eq. (5.69). These three losses are considered as the heat source for the ensuing thermal analysis. Then, a steady-state thermal analysis is performed to calculate the inductor temperature distribution. The analysis result is presented in Fig. 5.45, where the maximum temperature (302.1 K) appears at the center of the inductor.

5.2.3 Design Optimization of 2-D Inductor

Extending the prior analysis, a two-dimensional inductor is designed using the structural topology optimization method. For the performance analysis of the inductor, Maxwell's equations are modified for a low-frequency time-harmonic analysis and solved using the finite element method; see Sect. 3.5 for the formulated governing

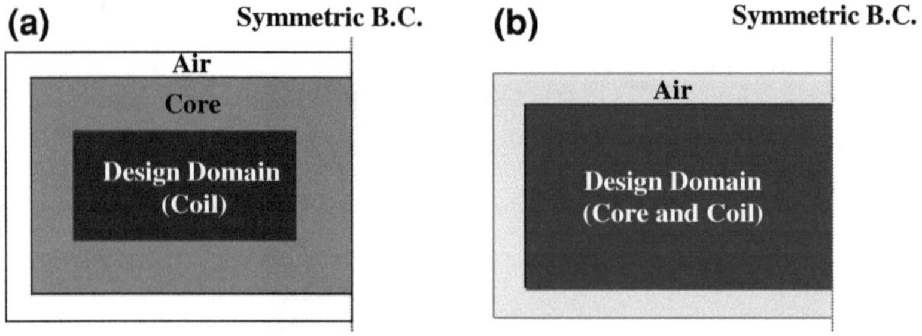

Fig. 5.46 Two-dimensional inductor design domain: **a** Example 1—coil only designed, **b** Example 2—coil and core designed simultaneously

equations. Here, two different examples are presented for the design optimization of inductors.

In the first example, the inductor coil shape is designed using a topology optimization approach with a fixed core shape, as shown in Fig. 5.46a. The second design example then aims to find the optimal coil and core shape of an inductor simultaneously. The design domain for the second example is shown in Fig. 5.46b. The design objective, i.e., to maximize the inductance, L, is the same in both numerical examples, where the inductance is calculated from the magnetic field distribution using Eq. (5.68). Thus, the objective function may be expressed as

$$\mathsf{F}_o = \frac{1}{\mu i^2} \int B^2 \mathrm{d}v, \tag{5.73}$$

and consequently, the optimization problems for the two examples are, respectively, summarized as follows.

For Example 1, where the coil only is designed

$$\begin{array}{rl} \text{Find} & \gamma \\ \text{Minimize} & \text{Eq. (5.73)} \\ \text{Subject to} & \text{Eqs. (3.34)} - \text{(3.38)} \\ & \int\limits_{\Omega_d} \gamma \mathrm{d}\Omega_d - v_u \leq 0 \\ & 0 \leq \gamma \leq 1 \\ \text{Given} & \varsigma = \varsigma_{\text{copper}} \gamma^p, \end{array}$$

and for Example 2, where the coil and core are designed simultaneously

Find	γ_1 and γ_2
Minimize	Eq. (5.73)
Subject to	Eqs. (3.34) – (3.38)
	$\int_{\Omega_d} \gamma_1 \mathrm{d}\Omega_d - v_1 \leq 0, \quad \int_{\Omega_d} \gamma_2 \mathrm{d}\Omega_d - v_2 \leq 0$
	$0 \leq \gamma_1 \leq 1, \quad 0 \leq \gamma_2 \leq 1$
Given	$\varsigma = \varsigma_{\text{copper}} \gamma_1^p, \quad \mu_r = \mu_{r\_\text{steel}} \gamma_2^p.$

In the latter optimization problem statement, γ_1 and γ_2 are design variables assigned to each finite element, ς is the electric conductivity, and μ_r is the relative magnetic permeability. Here, the coil volume of the second optimization problem, that is v_2, is fixed to 10 % of the whole design domain.

The design results for the first example are presented in Fig. 5.47a– c, where the full structure is shown by reflecting the results about the model symmetry line. Here, the relative permeability and electric conductivity of the core material are set as 150 and 1 S/m, respectively, and the electric conductivity of the copper is set to 5.96×10^7 S/m. The design optimization is performed for three different operating frequencies including 1 kHz, 5 kHz, and 10 kHz. The numerical model for this example, as well as the following example, was discretized using 8,800 linear quadrilateral elements, and each iteration for the optimization loop required 3 s to solve on a quad-core workstation with a 3.4 GHz processor and 16 GB of RAM. As can be seen in the design results in Fig. 5.47, the coil shape becomes thinner as frequency increases. This result may be attributed to fact that the skin effect is more prominent as frequency increases. Thus, several thin wires are more effective than fewer thick wires in increasing the total current flowing through the coil, and this consequently maximizes the inductance.

Finally, the design results for the second example are presented in Fig. 5.48. In this case, the operating frequency is fixed at 500 kHz, and the relative permeability and electric conductivity of the core material are again, respectively, set as 150 and 1 S/m. The optimization process is performed for different core volume constraint values leading to the design results shown in Fig. 5.48. A circular shape is obtained as the optimal coil design regardless of the core volume fraction. A rounded shape is also obtained for the optimal core designs. As the volume of the core material is increased from 40 to 90 %, the inductance, L, increases from 14.5 mH to 36.3 mH. However, the inductance does not increase much after reaching an 80 % core volume fraction. This implies that the design engineer might be able to reduce the core volume without losing the inductance after reaching a specific core volume percentage, and the additional core might not be necessary for the magnetic field distribution.

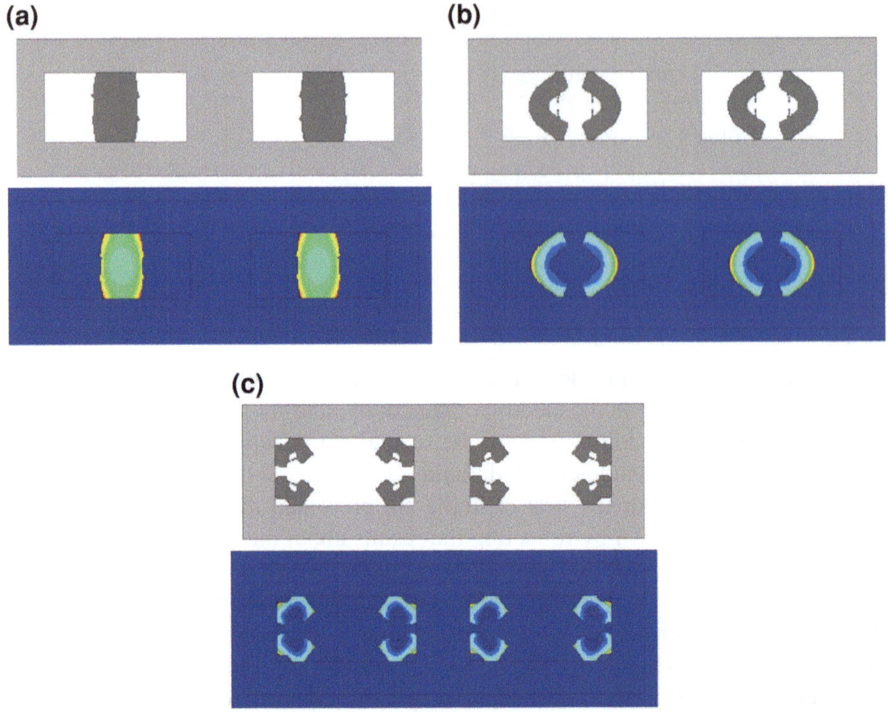

Fig. 5.47 Design Example 1, inductor coil design results (*top row*) and current density distribution (*bottom row*) for each operating frequency: **a** 1 kHz, **b** 5 kHz, and **c** 10 kHz. Note that for the design results shown in the *top row*, the *dark gray regions* indicate the coil conductor material, *while light gray regions* indicate the core material. For the current density distribution shown in the *bottom row*, *red-colored contours* indicate a high current density and *blue-colored contours* indicate a low current density

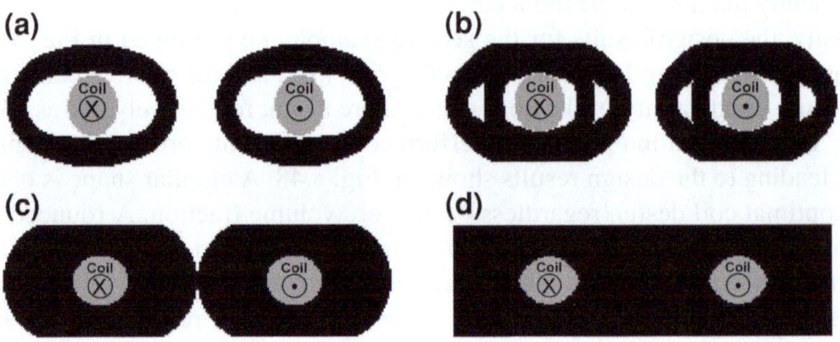

Fig. 5.48 Design Example 2, inductor core and coil design results: **a** 40 % core volume (L = 14.5 mH), **b** 60 % core volume (L = 27.4 mH), **c** 80 % core volume (L = 36.0 mH), and **d** 90 % core volume (L = 36.3 mH). Note that the *light gray regions* indicate the coil conductor material, *while dark regions* indicate the core material

5.3 RF Component Analysis and Design

Radio frequency components, or RF components, represent a group of devices which control and utilize radio waves. When the frequency of the AC current increases, a wave composed of oscillating electric and magnetic fields is radiated from the electromagnetic source. This phenomenon is referred to as an electromagnetic wave. The electromagnetic wave propagates though space at the speed of light regardless of its wavelength. Electromagnetic waves with wavelengths longer than those for infrared light are called radio waves.

Topology optimization was originally developed for structural mechanics problems and has been recently adapted for electromagnetic wave propagation problems including RF device design.

The history of topology optimization for electromagnetic wave propagation problems may be traced back to problems involving insulating materials even though most RF device design is based on the use of conductive materials. This is due to the numerical difficulties associated with the parametrization of conductive materials in high frequency electromagnetic analysis.

Photonic waveguides have been designed by considering 2-D scalar wave propagation in optical waveguides using FEM [47, 91]. Kiziltas, et al. [53] developed an optimum design method for the substrate of patch antennas using a 3-D Finite Element-Boundary Integration method (FE-BI). Nomura et al. [74] introduced a 3-D broadband topology optimization method for dielectric antennas using an adjoint FDTD technique [71].

For conductive materials, many structural optimization methods have been proposed based on gradient or meta-heuristic methods, such as genetic algorithms (GA) [49] or particle swarm optimization (PSO) [48]. Generally, most gradient-based methods may lead to local optima, while most meta-heuristic methods are extremely demanding in terms of computational cost even for a few design degrees of freedom. Therefore, a capable topology optimization method for conductive materials has been expected for a long time.

Topology optimization for conductive materials has been proposed for the optics domain [3] using the complex permittivity as a design variable. Since the absolute value of the permittivity of metals is relatively small in such frequency domain, this shrinks the difference between the skin depth and wavelength. However, when it comes to the RF domain, the difference between the skin depth and wavelength becomes severe and prevents the application of topology optimization due to the difficulty of interpolating the conductivity between an insulator and a good conductor [33]. One workaround is to represent such conductors using boundary conditions. The perfect electric conductor, PEC, boundary condition is the simplest representation of a conductor in that it assumes infinite conductivity. Similarly, an impedance boundary condition is often used to represent metallic surfaces with finite conductivity. The impedance boundary condition is usually applicable only for outermost boundaries. Aage et al. proposed a novel approach for this problem by embedding an impedance boundary condition in all facets of the mesh in the design domain for designing bulk conductive structures [1].

In this section, we introduce two design studies for RF devices. The first study pertains to a microstrip line device design, which involves conductive material design with a frequency domain analysis. The second study then focuses on a dielectric resonator antenna (DRA) with a time domain analysis plus a multiphysics extension.

5.3.1 Design Optimization of Microstrip Device

This section presents a structural optimization method for microstrip components based on topology optimization [75]. A microstrip is a type of transmission line for RF signals built on a printed circuit board, PCB. Sometimes, a microstrip device is constructed using a multilayer PCB. In the lower frequency range, interconnection between layers is simple since a metalized through-hole connection, or so-called electrical via, can be utilized. However, when the operational frequency is higher, usage of vias is problematic in terms of loss, and performance depends on the quality of the via fabrication. An RF coupler can be thought of as a higher frequency equivalent to an electrical via, where the RF signal is transmitted from one layer of a PCB structure to another layer without the physical (i.e., via) connection. The elimination of the via allows for the reduction of the associated parasitics, and hence, enables a more effective RF device. Thus, in this study, a microstrip coupler is optimized using a model that incorporates appropriate boundary conditions for the utilized numerical method, i.e., FEM. By locally controlling the parameter of the boundary conditions, it is possible to represent any configuration in the given design domain. The optimization utilizes sensitivity information obtained through adjoint analysis; refer to Sect. 4.1.

Design Variables

Consider the problem of determining the boundary of a design domain, Ω_d, by minimizing or maximizing an objective function. Following the standard topology optimization formalism, we introduce a fixed extended design domain, D, that includes the original design domain, Ω_d, and defines the following characteristic function, χ_{Ω_d}:

$$\chi_{\Omega_d}(\mathbf{x}) = \begin{cases} 1 & \text{if } \mathbf{x} \in \Omega_d \\ 0 & \text{if } \mathbf{x} \in D \setminus \Omega_d \end{cases}, \tag{5.74}$$

where $\mathbf{x}$ denotes a position in the extended design domain, D.

In this study, we apply a technique based on the density method outline in Chap. 4 for representing the conductive structure. Here, we assume that the design domain is composed of isotropic conductive materials, and the electric property determining the performance is electrical conductivity. We also use a logarithmic linear interpolation for the density function, where the relative conductivity at point, $\mathbf{x}$, can be expressed as

$$\exp\Big(\ln(\varsigma_{\mathrm{I}})\,(1-\gamma(\mathbf{x})) + \ln(\varsigma_{\mathrm{M}})\gamma(\mathbf{x})\Big), \quad (0 \leq \gamma(\mathbf{x}) \leq 1), \tag{5.75}$$

where ς_{I} is the electrical conductivity of the insulator material, and ς_{M} is that of the conductor material. This equation is applicable for boundary conditions that use conductivity as the material property.

The impedance boundary condition is mostly used to model a conductive structure with finite conductivity; however, this boundary condition can only be used on the outermost boundary of the analysis domain for truncation. This is because the electromagnetic field inside of the conductor is very small due to the skin effect, and thus, the field is not solved for, so the domain is excluded. This means that using the impedance boundary condition for topology design parameterization involves changes to the design domain and analysis domain. This makes the topology optimization problem very complicated. So, in this case topology optimization, *per se*, must be abandoned in lieu of shape optimization.

On the other hand, the transition boundary condition (TBC) is another boundary condition for modeling a finite conductive structure. The difference between this boundary condition and the previous impedance boundary condition is that the TBC assumes penetration of the electromagnetic fields across the boundary. The TBC models thin conductive surfaces by enforcing the following relationship:

$$Z_s \hat{\mathbf{n}} \times (\mathbf{H}_1 - \mathbf{H}_2) = \hat{\mathbf{n}} \times \left(\hat{\mathbf{n}} \times \mathbf{E}\right), \tag{5.76}$$

where Z_s is the surface impedance of the boundary, and $\mathbf{H}_1$ and $\mathbf{H}_2$ are, respectively, the magnetic field on one side and the other side of the boundary. The above equation balances the contribution of continuity across the boundary and conduction on the surface via Z_s. This means, this boundary condition is capable of representing both conductor and insulator by changing the value of Z_s as follows:

$$\begin{cases} \hat{\mathbf{n}} \times (\mathbf{H}_1 - \mathbf{H}_2) = 0 & (Z_s \to \infty) \quad \text{Insulator} \\ 0 = \hat{\mathbf{n}} \times \left(\hat{\mathbf{n}} \times \mathbf{E}\right) & (Z_s \to 0). \quad \text{Conductor} \end{cases} \tag{5.77}$$

Similar to the conductivity, Z_s is interpolated by γ

$$\exp\Big(\ln(Z_{s\_\mathrm{I}})\,(1 - \gamma(\mathbf{x})) + \ln(Z_{s\_\mathrm{M}})\gamma(\mathbf{x})\Big), \quad (0 \leq \gamma(\mathbf{x}) \leq 1), \tag{5.78}$$

where the appended subscripts $_{s\_\mathrm{I}}$ and $_{s\_\mathrm{M}}$, respectively, indicate values for an insulator and conductor.

Formulation of Optimization Problem

Figure 5.49 shows one example of an analysis model used for designing a microstrip patch antenna. As shown in this figure, the analysis domain contains a rectangular box, PCB substrate, and planar square area, D, that is placed on the top side of the box. The analysis domain is truncated by scattering boundaries, and the substrate is suspended at the center. The bottom of the substrate is defined as a conductive surface to represent the ground plane. On the top side of the substrate, a microstrip line and square, D, is drawn. The input port is on the edge of the substrate.

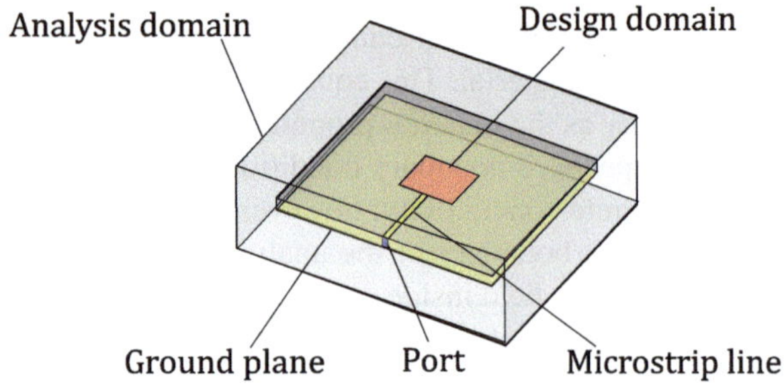

Fig. 5.49 An example microstrip antenna analysis model

Scattering parameters can be utilized as the objective function with frequency domain solvers. For maximizing radiation at n frequencies, a possible objective function is given by

$$\mathsf{F}_o = \sum_{i=1}^{n} w_i \frac{1}{2} |S_{11}(\omega_i)|^2, \tag{5.79}$$

where $|S_{11}(\omega_i)|$ is the return loss at frequency point i, and w_i is the weighting value for each frequency point.

On the other hand, for designs dealing with power distribution, it is natural to use the Poynting vector as the objective function

$$\mathsf{F}_o = \sum_{i=1}^{n} w_i \hat{\mathbf{n}}_i \cdot \mathrm{Re}\left(\mathbf{E} \times \mathbf{H}^*\right), \tag{5.80}$$

where $\hat{\mathbf{n}}_i$ is the normal unit vector of i-th evaluation surface.

Using traditional topology optimization, the optimization problem is formulated as follows:

Find γ
Minimize Eq. (5.79) or (5.80)
Subject to Maxwell's equations; refer to Chap. 3, Sect. 3.6
$0 \leq \gamma \leq 1$
Given Eq. (5.75) or (5.78).

Or, in the case that smooth geometry is preferred, a Heaviside projection-based formulation can be applied.

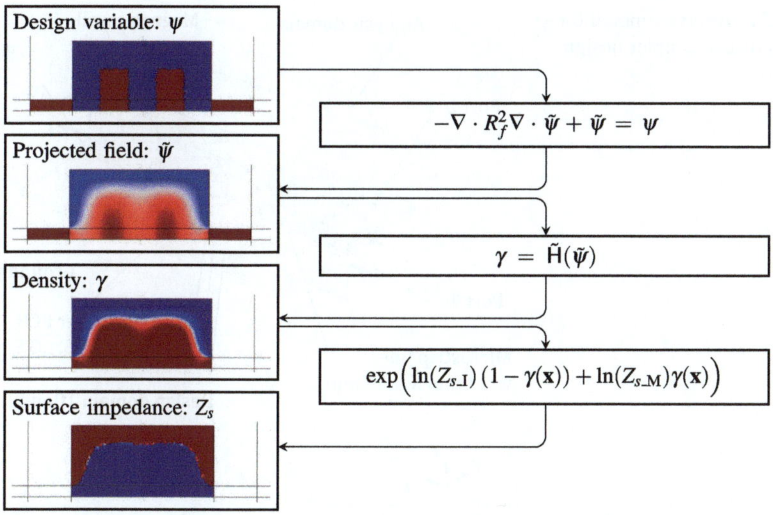

Fig. 5.50 Projection path from design variable to surface impedance for the microstrip optimization problem

$$
\begin{aligned}
&\text{Find} \quad \gamma \\
&\text{Minimize Eq. (5.79) or (5.80)} \\
&\text{Subject to Maxwell's equations; refer to Chap. 3, Sect. 3.6} \\
&\qquad \gamma = \tilde{\mathsf{H}}(\tilde{\psi}) \\
&\qquad -R_f^2 \nabla^2 \tilde{\psi} + \tilde{\psi} = \psi \\
&\qquad -1 \leq \psi \leq 1 \\
&\text{Given Eq. (5.75) or (5.78),}
\end{aligned}
$$

where R_f is the filter radius for Helmholtz filtering [51], as explained in Sect. 4.1 of Chap. 4. The projection and interpolation steps from design variable to surface impedance are summarized in Fig. 5.50.

As previously described, this optimization problem can be solved by any type of mathematical programming method. In this study, gradient-based non-linear programming methods, i.e., Sequential Linear Programming, SLP [96], and Method of Moving Asymptotes, MMA [90], are used with design sensitivity information that is obtained efficiently by the adjoint method.

During the optimization procedure, the design variable may contain intermediate values, or so-called grayscale. This is important to give slackness to the optimization problem; however, such values should be eliminated in the final solution. In the formulation above, it is possible to control the amount of the grayscale by tightening the transition bandwidth, h_t, in the relaxed Heaviside function Eq. (4.9). Here, h_t is

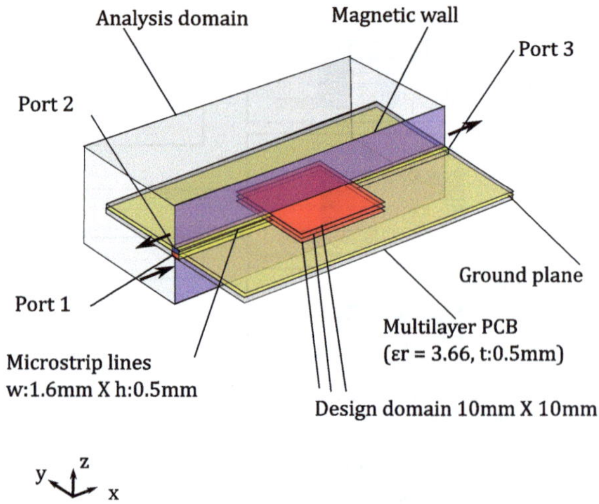

Fig. 5.51 Analysis model for the microstrip coupler design problem

controlled by the following formula:

$$h_{t\_\max}\left(h_{t\_\min}+\left(1-\frac{n^p}{n^p+w^p}\right)(1-h_{t\_\min})\right), \tag{5.81}$$

where $h_{t\_\min}$ and $h_{t\_\max}$ are, respectively, the minimum and maximum values for h_t, n is the current iteration step, w is the half bandwidth of transition in iteration steps, and p is the control parameter of the curve profile.

Device Setup

A microstrip coupler for vertical interconnect on a multilayer PCB is designed using a Heaviside projection formulation combined with MMA [90], as described above. Figure 5.51 shows a schematic of the analysis model used for the coupler design problem. The coupler is designed to connect microstrip lines on different layers of a multilayer PCB. The PCB has three metal layers and two 0.5 mm thick dielectric substrates between the metal layers. The input microstrip line, Line 1, is on the bottom metal layer and connects Port 1 to the design domain. The design domain consists of 10 mm squares on each metal layer. The middle metal layer is used as the ground plane. On the top metal layer, the output microstrip lines, Lines 2 and 3, are connected from the design domain and terminated by Ports 2 and 3, respectively. All microstrip lines are 1.6 mm in width.

The Poynting vector-based formulation, Eq. (5.80), is used for the objective function. Each Poynting vector is calculated at the cross-section of each microstrip line at the middle of the line. The weighting value is 0.1, −0.45, and 0.45 for Line 1, 2 and 3, respectively. The analysis frequency is 4.0 GHz.

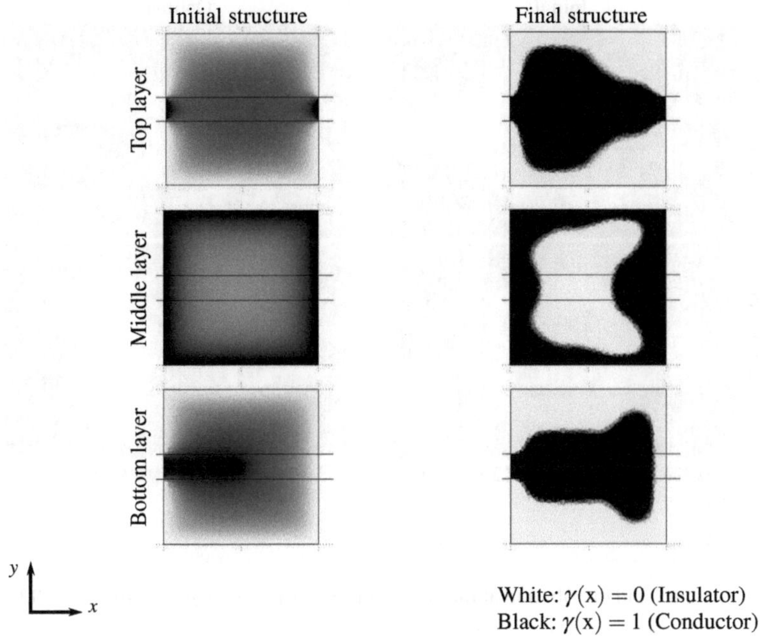

Fig. 5.52 Density distribution of the initial and final microstrip coupler structure

Optimization Results

The optimization starts from a very trivial initial design, as shown in the left-hand side column of Fig. 5.52. The value of the initial design variable, ψ_{init}, is all zero, i.e., $\gamma = 0.5$. However, in order to avoid electric field concentration along the border of the design domain, the value of γ on the boundaries is controlled to be a specified value such that $\gamma = 0$ on insulative area boundaries and unity on conductive area boundaries. This is done by enforcing a Dirichlet boundary condition of the projection equation (i.e., scalar Helmholtz equation).

Figure 5.52 shows the initial and optimized density distribution on each layer. The initial design just shows a vague distribution gradually varying from either metal or insulator on the constrained boundary to intermediate material. On the other hand, the optimized result shows a clear black and white topology with a very narrow transition area from metal to insulator. Figure 5.53 shows the computed electric field distribution at the middle of the lower and upper substrates of the device at 4.0 GHz.

Fabrication and Experiments

A multilayer printed circuit board fabrication technique is used for preparing a measurement sample. Since the design obtained by topology optimization is a scalar field, it has to be converted into a vectorial drawing that is compatible with CAD-based fabrication systems. As observed in Fig. 5.52, the obtained design has a relatively clear boundary between material and void regions, thanks to the transition bandwidth

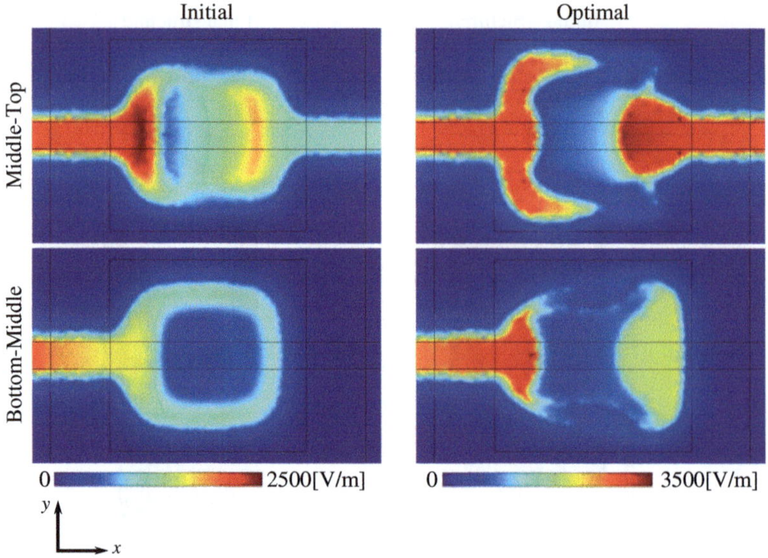

Fig. 5.53 Electric field distribution at the *middle* of the *lower* and *upper* substrates at 4.0 GHz

control. Thus, the vectorial data can be extracted using ordinary contour extraction algorithms. The extracted device CAD geometry is illustrated in Fig. 5.54 and a representative prototype fabricated using this geometry is shown in Fig. 5.55.

To confirm the performance of the final device design, the response of the coupler was predicted via a follow-up simulation using the synthesized CAD geometry. In this model, the pattern is defined geometrically with spline curves. The microstrips are extended with bends to have Port 1 and Port 2 at different positions on an edge for attaching SMA connectors for measurements. The ground plane is also added on the top and bottom side of the multilayer substrate.

The computed response of the microstrip coupler device is shown in Fig. 5.56. The power level of the top layer is one-half of the power level on the bottom layer, and power is equally distributed to the right and left on the top layer. The bottom row of Fig. 5.56 shows a zoomed view of the coupler region; the field distribution is almost the same as that shown in Fig. 5.53. Additionally, Fig. 5.57 shows the simulated and experimentally measured S-parameters of the generated coupler device. The RF microstrip coupler design has a reasonable working bandwidth with equal power distribution, and a decent match is obtained between the simulation and experiment.

In summary, we proposed a flexible and efficient structural optimization method for microstrip components in this section. The method is based on topology optimization with adjoint analysis. The microstrip configuration is modeled by the conductivity of the surface. Numerical examples and representative prototypes were described to demonstrate the practicality of the method.

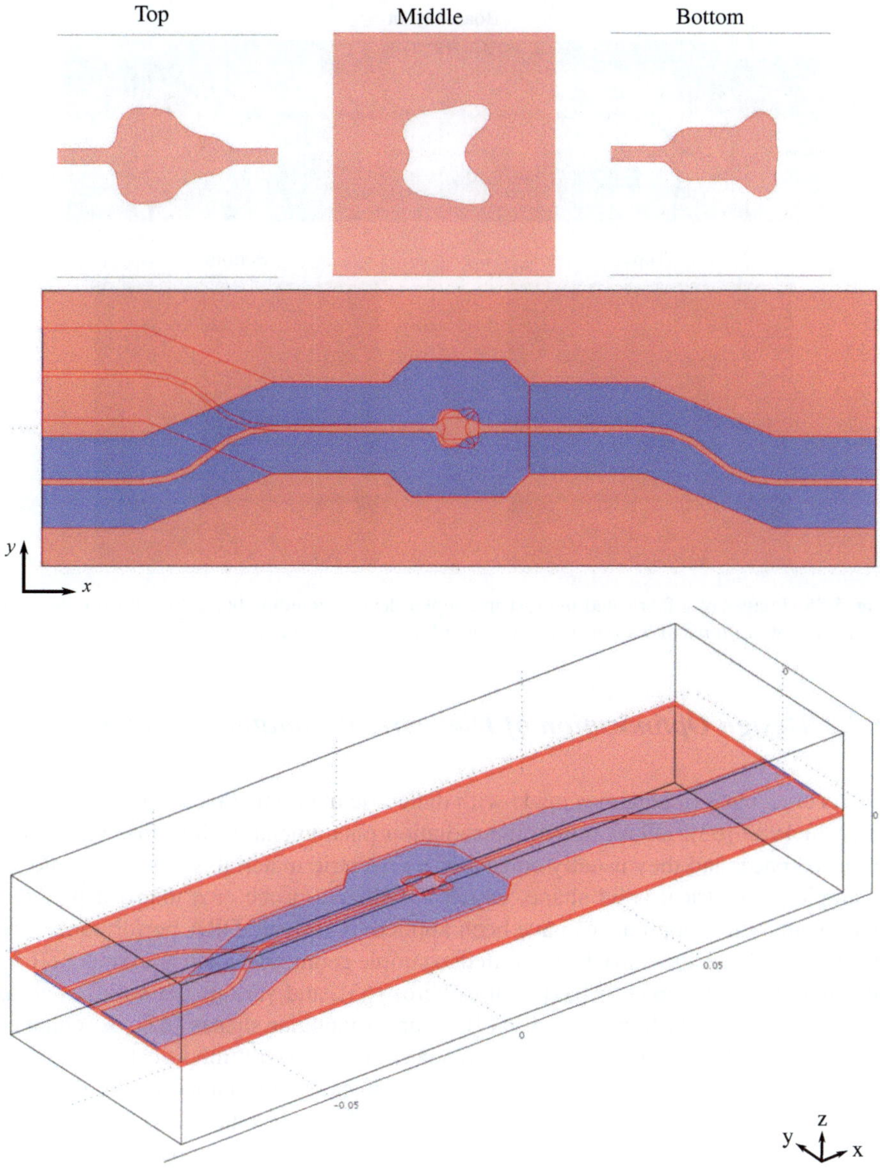

Fig. 5.54 Extracted vector outlines of the various substrate layers (*top row*) and real microstrip coupler device CAD model (*middle row*: top view, *bottom row*: perspective view)

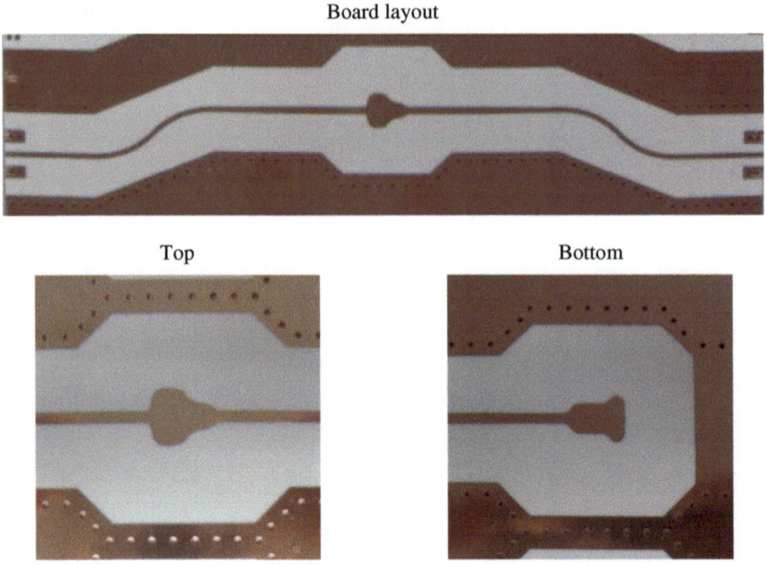

Fig. 5.55 Images of a fabricated microstrip coupler device including board layout (*top row*) and zoomed views of *top* and *bottom* side of the board in a local coupler region

5.3.2 Design Optimization of Dielectric Resonator Antenna

DRAs are a class of antennas made with dielectric materials first proposed by Long et al. in 1983 [63]. DRAs utilize the radiation phenomena of dielectric resonators in open space, and they usually consist of a dielectric material, such as ceramic, in a simple geometrical solid shape, which is fed by a probe or a microstrip. Since mainly analytical approaches have been applied to evaluate DRA performance, the proposed DRA shapes have been limited to simple geometries such as cylinders [63], rectangular parallelepipeds [64], hemispheres [65], and various flat solids such as rectangular discs [68], circular discs [61], and triangular shapes [62]. To enhance the operational bandwidth of these antenna designs, several modifications of simple DRA shapes have been proposed, such as stacking resonators having different permittivity [52], combining resonators having different resonance frequencies [35], introducing an insert or an air-gap between the ground plane and the resonator [18], making a cylindrical resonator annular [67], and providing a rectangular resonator with notches [82].

Topology optimization is usually solved using the finite element method since it is mathematically well established. Meanwhile, for engineering applications involving high frequency electromagnetics, different numerical analysis methods are frequently used. Time domain methods are one group of such methods. Usually, the finite element method is combined with a frequency domain formulation, as discussed in Sect. 3.6, whereas the time domain method literary solves Maxwell's equations

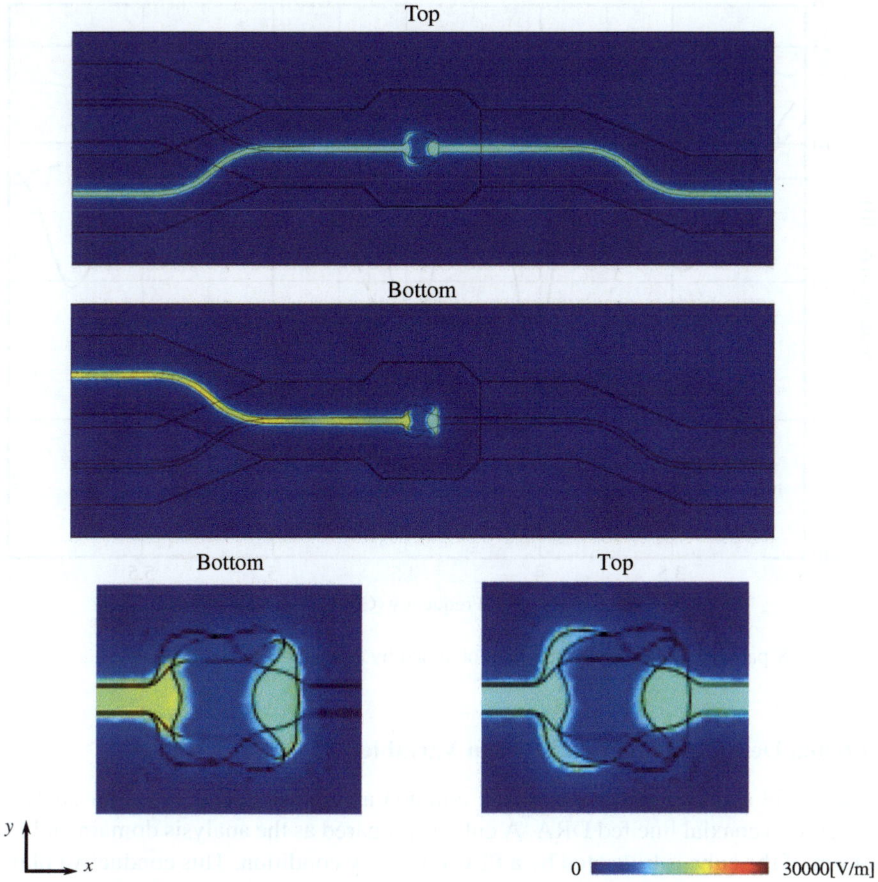

Fig. 5.56 CAD model simulation of electric field distribution at the *middle* of the *lower* and *upper* substrates at 4.0 GHz

in a time domain formulation by discretizing time, t. The major benefit of a time domain analysis is its ability in broadband analysis. By inputting a pulse that contains broadband frequency components, the time domain analysis allows for the solution of the frequency response for all pulse components with a single analysis; this is enabled by the well-known Fourier transform. The finite difference-time domain method, FDTD [94] method, is numerically robust and can handle complex general models that include an arbitrary number of multiple media, such as dielectrics, magnetic devices, and lossy materials. Such features make this method the most popular approach to evaluating antenna performance.

This section provides some example studies of DRAs using FDTD, and offers a multiphysics extension with consideration of fabrication constraints.

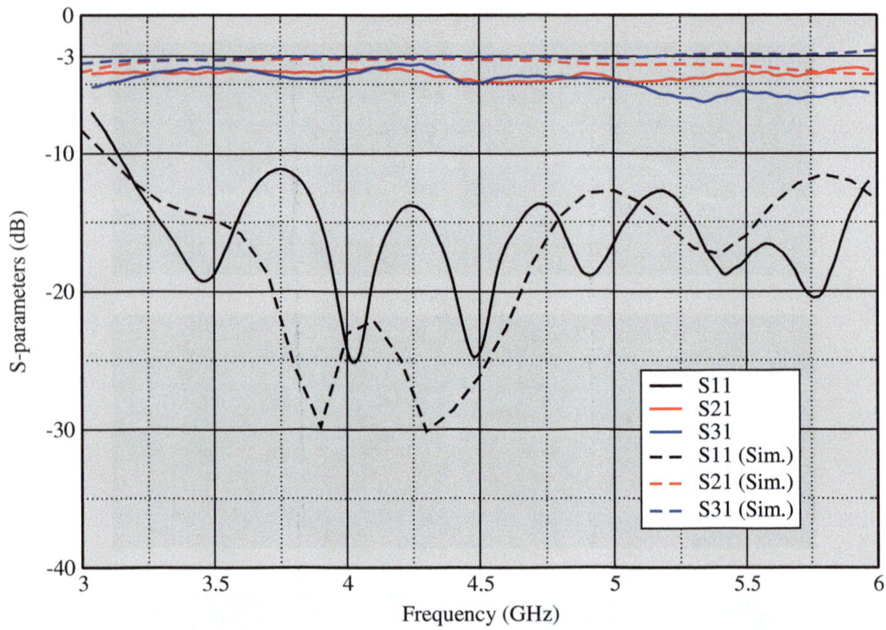

Fig. 5.57 S-parameters of coupler structure obtained by simulation and experiment

Antenna Design Problem and Design Variable

Figure 5.58 shows a schematic of the antenna analysis model. This is an analysis model for a coaxial line fed DRA. A cube is prepared as the analysis domain and the bottom of the cube is truncated by a PEC boundary condition. This conductive plane works as the ground plane. The other sides are truncated by absorbing boundaries, which are often referred to as perfectly matched layers, PMLs [7]. At the center of the ground plane, a rectangular box is prepared as the fixed design domain. Beneath the fixed design domain, a coaxial cable is connected from the bottom. The end of the cable is a port to feed electromagnetic energy to the system.

Here, we set up the structural design problem, where similar to Sect. 5.3.1, we consider the design problem of determining the boundary of a design domain, Ω_d, by minimizing or maximizing an objective function. Once more, the key idea of the topology optimization method is the introduction of a fixed, extended design domain, D, that includes the original design domain Ω_d, *a priori*, and the utilization of the following characteristic function χ_{Ω_d}, per Eq. (5.74)

In this study, we again apply the SIMP method [5] to the design of DRAs. Here, we assume that the design domain is composed of a dielectric material and air, and the physical property determining the antenna's performance is the distribution of electrical permittivity. The relative permittivity at point $\mathbf{x}$ is expressed as

$$\epsilon(\gamma(\mathbf{x})) = \epsilon_{\text{air}} + (\epsilon_{\text{solid}} - \epsilon_{\text{air}})\gamma(\mathbf{x})^p, \quad (0 \leq \gamma(\mathbf{x}) \leq 1), \tag{5.82}$$

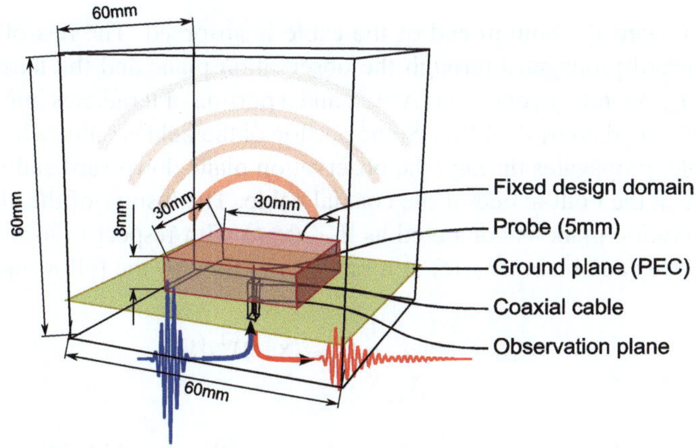

Fig. 5.58 Analysis model of dielectric resonator antenna

where ϵ_{air} and ϵ_{solid} are the permittivity in air and the solid dielectric material, respectively. While the SIMP method introduces a penalization parameter, p, to penalize intermediate design values by modifying the relationship between weight and stiffness for structural mechanics problems, here we use a simple linear interpolation by setting p to 1.0, since permittivity is considered to change linearly with respect to the normalized density, $\gamma(\mathbf{x})$. In the presented numerical examples, we confirm that improvements in the optimal configurations using different values of p are of mixed utility and not entirely practical, when compared with the case where p is set to 1.0.

Formulation of Optimization Problem

Here, the reflection coefficient $|S_{11}(\omega)|$, i.e., the return loss is formulated by the following:

$$|S_{11}(\omega)| = \left| \frac{\mathscr{P}_{\text{ref}}(\omega)}{\mathscr{P}_{\text{in}}(\omega)} \right|, \tag{5.83}$$

which is a measure of the radiation efficiency for the electromagnetic waves radiating from the design domain, where $\mathscr{P}_{\text{in}}(\omega)$ and $\mathscr{P}_{\text{ref}}(\omega)$ are the input power and reflected power in the frequency domain, respectively. That is, by minimizing $|S_{11}(\omega)|$, the radiated power, $\mathscr{P}_{\text{rad}}(\omega)$, which is a measure of the radiation performance of the antenna, is maximized, since we are operating under the lossless assumption formulated as

$$\mathscr{P}_{\text{in}}(\omega) = \mathscr{P}_{\text{rad}}(\omega) + \mathscr{P}_{\text{ref}}(\omega). \tag{5.84}$$

Note that, Eq. (5.83) is most commonly used as a performance measure for antenna development.

To calculate $\mathscr{P}_{\text{ref}}(\omega)$, an electric field in the form of a Gaussian pulse is fed in at the feeding plane in transverse electric and magnetic (TEM) mode. This electric field propagates in both upward and downward directions, but the power directed downward toward the bottom end of the cable is absorbed. The rest of the power directed upward propagates through the observation plane and the inactivated top end absorbing boundary condition (ABC), and a portion of it radiates into the air and is ultimately absorbed by the PML. Some portion of the field is reflected, reenters the coaxial cable, propagates through the observation plane downward and is absorbed by the ABC at the bottom end of the coaxial cable. The history of the electric field at the observation plane is computed as $\mathbb{E}_{obs}(\mathbf{x}, t)$ with respect to location, $\mathbf{x}$, and time, t, and we can obtain $\mathscr{P}_{\text{ref}}(t)$ as a function of time by the following equation:

$$\mathscr{P}_{\text{ref}}(t) = \int_{\Omega_m} \frac{1}{2} (\mathbb{E}_{\text{ref}}(\mathbf{x}, t))^2 \, \mathrm{d}\Omega_m, \tag{5.85}$$

where $\mathbb{E}_{\text{ref}}(\mathbf{x}, t)$ denotes the reflected component of the electric field at point $\mathbf{x}$, Ω_m is the area of observation plane and $\mathscr{P}_{\text{ref}}(\omega)$ is obtained by Fourier transformation as follows:

$$\mathscr{P}_{\text{ref}}(\omega) = \int_{-\infty}^{\infty} \mathscr{P}_{\text{ref}}(t) e^{-i\omega t} \mathrm{d}t, \tag{5.86}$$

and $\mathbb{E}_{\text{ref}}(\mathbf{x}, t)$ is calculated by

$$\mathbb{E}_{\text{ref}}(\mathbf{x}, t) = \mathbb{E}_{\text{obs}}(\mathbf{x}, t) - \mathbb{E}_{\text{in}}(\mathbf{x}, t), \tag{5.87}$$

where $\mathbb{E}_{\text{in}}(\mathbf{x}, t)$ is the electric field at the observation plane excited by the input pulse alone, without reflection, and is calculated separately.

Since $\mathscr{P}_{\text{in}}(\omega)$ is constant in this optimization problem, minimizing $|S_{11}(\omega)|$ is equivalent to minimizing $\mathscr{P}_{\text{ref}}(\omega)$. When the input pulse has a power spectrum with the same peak frequency as the antenna's working frequency, and has the required bandwidth, minimizing $\mathscr{P}_{\text{ref}}(t)$ subject to the chosen input pulse minimizes $\mathscr{P}_{\text{ref}}(\omega)$ in the specified bandwidth, with the input pulse weighted to have its peak at the working frequency. Thus, the objective function is formulated as an integral value of $\mathscr{P}_{\text{ref}}(t)$ with respect to time, as follows:

$$\mathsf{F}_o = \int_0^{t_f} \mathscr{P}_{\text{ref}}(t) \mathrm{d}t, \tag{5.88}$$

where t_f is a fixed final time, and a Gaussian pulse having a center frequency equal to the target frequency, and a −3 dB specified power bandwidth, is selected for the input pulse. Therefore, the optimization problem is formulated as described in Sect. 4.1, where the above objective function is minimized subject to a volume constraint, the material interpolation function Eq. (5.82), and Maxwell's equations, as discussed in Chap. 3.

Adjoint Analysis

Following the related discussion in Sect. 4.1, adjoint analysis can also be carried out in the time domain. First, the objective function expressed in Eq. (5.88) is rewritten as follows:

$$\mathsf{F}_o = \int_{\Omega_m} \int_{t=0}^{t_f} G(\bar{\mathbb{E}}, \gamma) \mathrm{d}t \mathrm{d}\Omega_m, \tag{5.89}$$

where $\bar{\mathbb{E}}$ is a discretized electric field vector,

$$G(\bar{\mathbb{E}}, \gamma) = \frac{1}{2} |\mathbf{E}_{\text{obs}}(t) - \mathbf{E}_{in}(t)|^2 \tag{5.90}$$

On the other hand, by discretizing Maxwell's equations in the time domain, the following equation can be obtained:

$$[M(\gamma)]\{\ddot{\bar{\mathbb{E}}}\} - [B(\gamma)]\{\dot{\bar{\mathbb{E}}}\} + [K(\gamma)]\{\bar{\mathbb{E}}\} = \{Q\}, \tag{5.91}$$

where dot denotes the derivative with respect to time, t. $[M(\gamma)]$, $[B(\gamma)]$ and $[K(\gamma)]$ are system matrices and $\{Q\}$ is the initial condition value for the time transient equation. Here, the residual, $\mathscr{R}(\mathbb{E})$, is introduced as follows:

$$\mathscr{R}(\mathbb{E}) = [M(\gamma)]\{\ddot{\bar{\mathbb{E}}}\} + [B(\gamma)]\{\dot{\bar{\mathbb{E}}}\} + [K(\gamma)]\{\bar{\mathbb{E}}\} - \{Q\}. \tag{5.92}$$

Using Eq. (5.91), F_o can be rewritten as F_o^* formulated below

$$\mathsf{F}_o^* = \int_{\Omega_m} \left[\int_0^{t_f} \left(G(\bar{\mathbb{E}}, \gamma) \right) \mathrm{d}t - \int_0^{t_f} \hat{\lambda}^{\mathrm{T}} \left(\mathscr{R} \right) \mathrm{d}t \right] \mathrm{d}\Omega_m, \tag{5.93}$$

where $\hat{\lambda}$ is the time-dependent adjoint variable vector. Then, taking the first variation of the above equation yields

$$\delta \mathsf{F}_o^* = \int_{\Omega_m} \left[\frac{\partial \mathsf{F}_o^*}{\partial \bar{\mathbb{E}}^{\mathrm{T}}} \{\delta \bar{\mathbb{E}}\} + \frac{\partial \mathsf{F}_o^*}{\partial \gamma^{\mathrm{T}}} \{\delta \gamma\} \right] \mathrm{d}\Omega_m. \tag{5.94}$$

By assuming $\hat{\lambda}(t_f) = 0$ and $\dot{\hat{\lambda}}(t_f) = 0$, we have the following adjoint equation in the time domain:

$$[M(\gamma)]\{\ddot{\hat{\lambda}}\} - [B(\gamma)]\{\dot{\hat{\lambda}}\} + [K(\gamma)]\{\hat{\lambda}\} = \left\{ \frac{\partial G}{\partial \bar{\mathbb{E}}} \right\}^{\mathrm{T}}, \tag{5.95}$$

where $\hat{\lambda}$ can be obtained by solving the above equation. The sensitivity vector can be calculated by substituting this value into Eq. (5.94). Equation (5.95) is equivalent to Eq. (5.91), except that the $\{\dot{\hat{\lambda}}\}$ term sign is negative, which implies that this is a terminal value problem, and $\left\{\frac{\partial G}{\partial \mathbb{E}}\right\}^{\mathrm{T}}$ is a terminal value of this time transient equation. Therefore, Eq. (5.95) can be solved using the same coefficient matrices by using the backward time direction scheme, with $\left\{\frac{\partial G}{\partial \mathbb{E}}\right\}^{\mathrm{T}}$ as the input current pulse in the observation domain of the objective function.

When the FDTD method is used in practical cases, these equations are approximated using finite difference grids in space and time. The design sensitivity of the objective function with respect to design variable, γ_i, is calculated as

$$\frac{\partial \mathsf{F}_o}{\partial \gamma_i} \approx -\int_{\Omega_{m\_i}}\int_0^{t_f}\left(\hat{\mathbb{E}}(t)^{\mathrm{T}} \cdot \frac{\partial \mathscr{R}(\bar{\mathbb{E}}(t))}{\partial \gamma_i}\right) \mathrm{d}t\mathrm{d}\Omega_m, \tag{5.96}$$

where $\bar{\mathbb{E}}(t)$ is the electric field obtained by the forward FDTD method, $\hat{\mathbb{E}}(t)$ is the adjoint variable with respect to $\bar{\mathbb{E}}(t)$, Ω_{m_i} is a domain of a Yee cell in which γ_i is located, and the bar over $\mathbb{E}$ is added to indicate that the values of $\bar{\mathbb{E}}$ have been previously calculated by forward FDTD, and are not variables.

Single Band Design

Here, several design cases of FDTD topology optimization are presented. Figures 5.59, 5.60, 5.61 and 5.62 show the optimized configurations (on left) and return loss of the DRAs (on right). Each configuration shows a bottom view of the antenna. For the return loss plots, the thin lines show the input pulse (green) and reflected pulse (red), while the thick lines show the return loss of the initial structure (green) and optimal structure (red). As seen in the return loss plots, the working frequency band (i.e., the frequency band where the return loss is less than -10 dB) broadly encompasses each target frequency.

The obtained structure shows clear separation of the void and material regions. This is achieved by a technique termed 'subcutoff pulse' [74]; refer to Fig. 5.63. Without any treatment, wideband optimization is likely to yield solutions exhibiting a large portion of grayscale area; however, the penalization scheme in SIMP does not work, since the objective function does not monotonically depend on the amount of material (in contrast with many structural design problems). However, if the target frequency is unfeasibly low, the objective function shows a monotonic relationship to the amount of material, because a larger resonator shows a lower resonant frequency, such that, the optimizer tries to fill all of the design domain with material. The subcutoff pulse technique utilizes this phenomenon. By mixing low frequency components into the input pulse, a monotonic problem is obtained. If the mixed frequency is below the cutoff frequency, it is infeasible. Additionally, since it is necessary to keep the void areas to reduce the objective at the target frequency

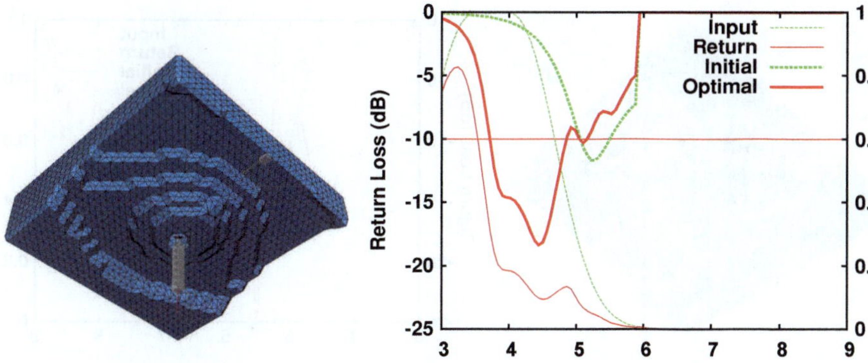

Fig. 5.59 DRA design targeted at 4.0 GHz. *Bottom view* of antenna shown (*on left*) with return loss plot (*on right*), where the *horizontal axis* indicates frequency in units of GHz

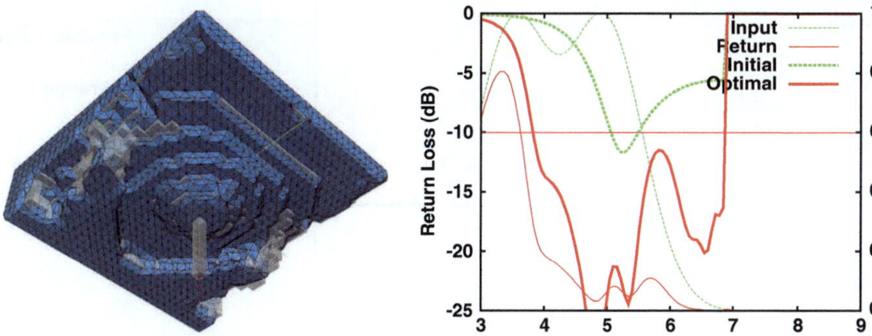

Fig. 5.60 DRA design targeted at 5.0 GHz. *Bottom view* of antenna shown (*on left*) with return loss plot (*on right*), where the *horizontal axis* indicates frequency in units of GHz

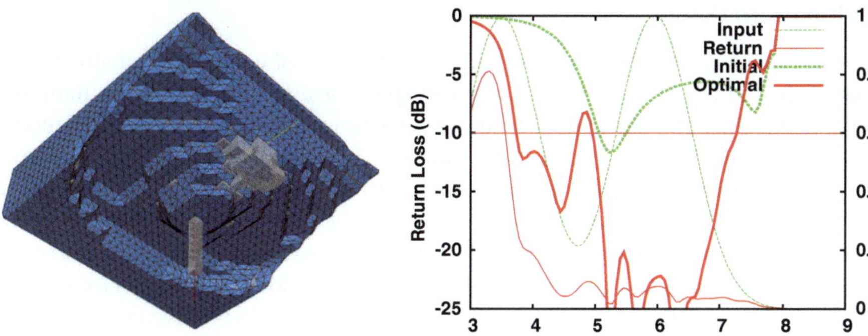

Fig. 5.61 DRA design targeted at 6.0 GHz. *Bottom view* of antenna shown (*on left*) with return loss plot (*on right*), where the *horizontal axis* indicates frequency in units of GHz

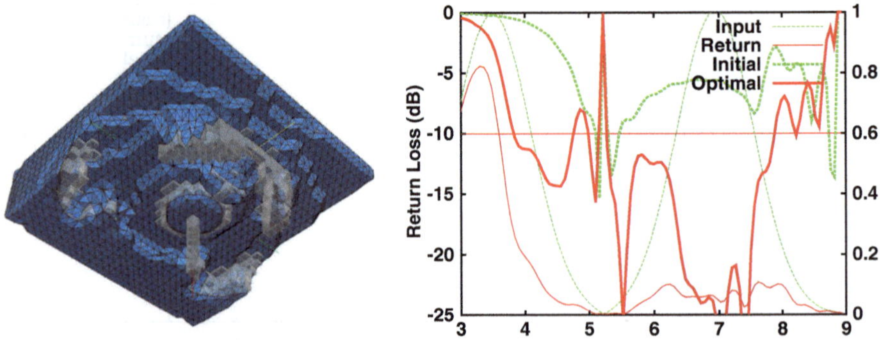

Fig. 5.62 DRA design targeted at 7.0 GHz. *Bottom view* of antenna shown (*on left*) with return loss plot (*on right*), where the *horizontal axis* indicates frequency in units of GHz

Fig. 5.63 Subcutoff pulse concept

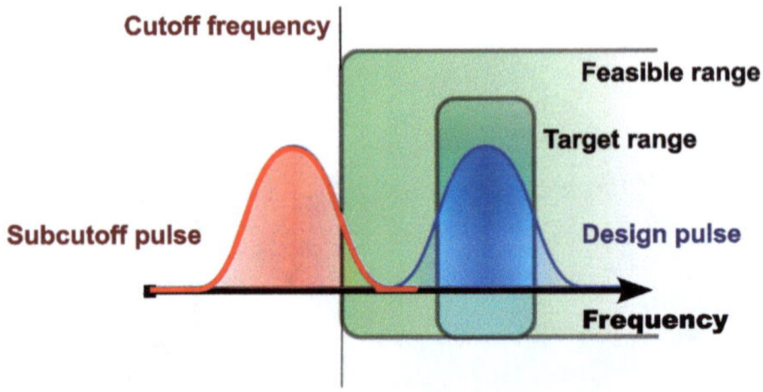

band, an implicit upper volume constraint effect arises. These two effects help in separating the void and material domains.

Multiphysics Design

As we see in Figs. 5.59, 5.60, 5.61 and 5.62, one side of each obtained structure is floating from the ground. This is not a very good design in terms of installation of the antenna since the gap distance can vary and sometimes antenna performance is sensitive to such distance. In order to avoid such situations, multiphysics design is an attractive solution. If the antenna structure is stiff enough in response to mechanical loads, it is possible to avoid installation issues, while improving the reliability of the ceramic part. For this purpose, the DRA design problem is solved in conjunction with a structural mechanics problem. Additionally, by doing so, we can expect that SIMP penalization becomes effective [73]. This may help to remove any remaining grayscale elements observed in Fig. 5.60 or 5.62.

The schematic for the structural mechanics setup is shown in Fig. 5.64. Assuming that the four corners of the bottom of the DRA are fixed, a vertical load is applied at the top center of the structure. The DRA is designed to have minimum compliance to show the maximum stiffness against the load.

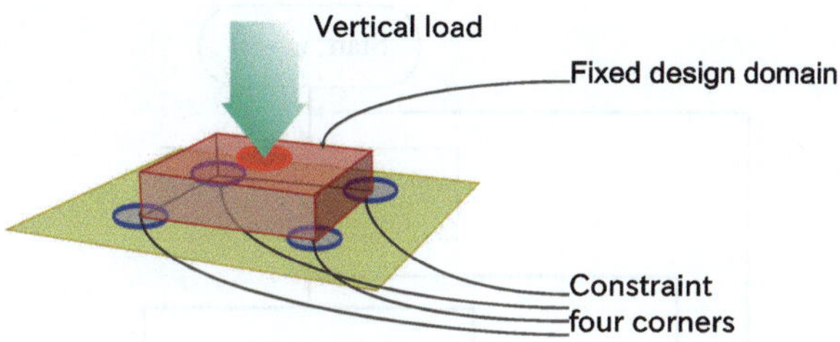

Fig. 5.64 Schematic of the structural problem setup for DRA stiffness improvement

The governing equation for a static structural problem in elastic equilibrium is

$$\int_D \delta\mathbf{u} : \mathbf{C} : \mathbf{u}\,d\Omega = \int_{\Gamma_t} \mathbf{t} \cdot \delta\mathbf{u}\,d\Gamma, \tag{5.97}$$

where $\mathbf{u}$ is the displacement vector, $\delta\mathbf{u}$ is virtual displacement vector, $\mathbf{C}$ is the elastic tensor, Γ_t is the boundary subjected to a surface load, and $\mathbf{t}$ is the surface load vector. The objective function for minimum compliance is then

$$l(\mathbf{u}) = \int_{\Gamma_t} \mathbf{t} \cdot \mathbf{u}\,d\Gamma. \tag{5.98}$$

Simultaneous optimization of antenna performance and structural stiffness involves multi-objective optimization. Here, by using a weighted sum approach with weighting coefficient, w, the optimization problem is formulated as follows:

$$\text{Find} \quad \gamma \tag{5.99a}$$

$$\underset{\gamma}{\text{Minimize}} \quad \mathsf{F}_o = w\frac{\mathcal{P}_{ref}}{\mathcal{P}_{ref}^{init}} + (1-w)\frac{l(\mathbf{u})}{l(\mathbf{u})^{init}} \tag{5.99b}$$

$$\text{Subject to} \quad 0 \leq \gamma \leq 1 \tag{5.99c}$$

$$v = \int_D \gamma\,d\Omega \leq v_u \tag{5.99d}$$

$$\text{Maxwell's equations; refer to Chap. 3, Sect. 3.6} \tag{5.99e}$$

$$\text{Eq. (5.97)} \tag{5.99f}$$

$$\text{Given} \quad \text{Eqs. (5.100)–(5.101)}, \tag{5.99g}$$

where w is the weighing coefficient, and the superscript, ^{init}, stands for an objective value with initial design, i.e., $\gamma = 0.5$ everywhere. The governing equations are coupled with the following material interpolation functions

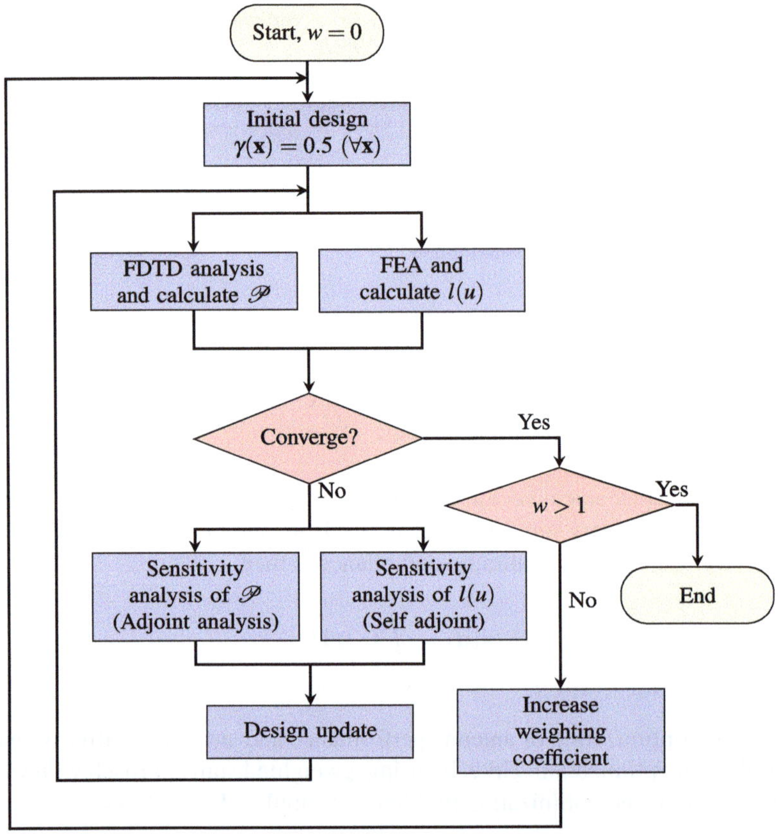

Fig. 5.65 Flowchart of Pareto front calculation for optimization problem, Eq. (5.99)

$$\epsilon(\gamma(\mathbf{x})) = \epsilon_{air} + (\epsilon_{solid} - \epsilon_{air})\gamma(\mathbf{x}) \tag{5.100}$$

$$\mathbf{C}(\gamma(\mathbf{x})) = \mathbf{C}_{solid}\gamma(\mathbf{x})^3. \tag{5.101}$$

Here, the upper bound of material volume is constrained as Eq. (5.99d). This is to utilize the penalization effect for grayscale values via a SIMP scheme. The SIMP scheme shows an effect when an upper volume constraint is combined with a penalty parameter value larger than unity. The penalty value is set to 3 as in Eq. (5.101). The upper volume bound is set to 0.5.

Since this setup uses SIMP for grayscale penalization, the subcutoff pulse technique is not used. Instead, a multi-band design problem is specified using pulse mixing. The input pulse is a summation of two sine modulated Gaussian pulses whose center frequencies are 5.0 GHz and 7.0 GHz.

Figure 5.65 shows a flowchart for the Pareto front calculation. Starting from $w = 0$, the multi-objective topology optimization problem is solved using both FDTD (for the electromagnetics analysis) and FEM (for the structural analysis).

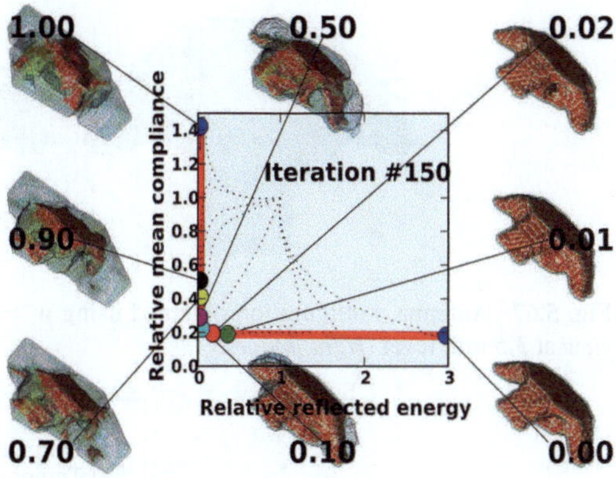

Fig. 5.66 Pareto front and corresponding configurations for multiphysics antenna optimization

With both numerical methods, state variables are solved, and objective functions for each physics and constraints are calculated. If the analyzed solution exhibits sufficient performance, the design iteration loop is terminated. Otherwise, a sensitivity analysis is performed. The electromagnetics problem involves adjoint analysis with time backward calculation, while the structural problem does not require additional analysis since it is a self adjoint problem. Using the obtained sensitivity information, the design variables are then updated using non-linear programming methods; in this study, SLP is used. When the solution is converged, the weighting coefficient is increased and the optimizer is restarted from a new blank initial design. When $w > 1$, all procedures are terminated.

Figure 5.66 shows the Pareto front obtained by the discussed optimization problem setup. The vertical axis shows the normalized structural objective value, $l(\mathbf{u})/l(\mathbf{u})^{init}$, and the horizontal axis shows the normalized electromagnetics objective value, $\mathscr{P}_{ref}/\mathscr{P}_{ref}^{init}$. The markers show the values for $w = 0.0$, 0.01, 0.02, 0.1, 0.5, 0.7, 0.9, and 1.0. The corresponding antenna design configuration is also shown in the figure. The red area indicates solid material, while transparent areas indicate grayscale regions. Dotted lines radiating from coordinate (1, 1) show the convergence path of each marker.

The Pareto front appears to be convex, considering the selected weighting values. With a weighting value less than 0.5, the design demonstrates desirable stiffness, and with w greater than or equal to 0.98, the solution shows good electromagnetics performance. However, in terms of grayscale penalization, designs with $w \geq 0.90$ are not sufficient. With precise analysis, we found that $w = 0.03$ is the solution that best balances the multiple objectives and grayscale penalization. Figure 5.67 shows the obtained antenna configuration with $w = 0.03$. The left side image shows the top view and the right side image shows a section view of a cut plane positioned at a distance of 3.5 mm from the base of the antenna. The obtained structure is a table-like structure with four legs and no grayscale exists in the solution. Figure 5.68 shows the frequency characteristics of the obtained design. The antenna has a sufficient, -10 dB, working bandwidth at both 5.0 GHz and 7.0 GHz, the two target center

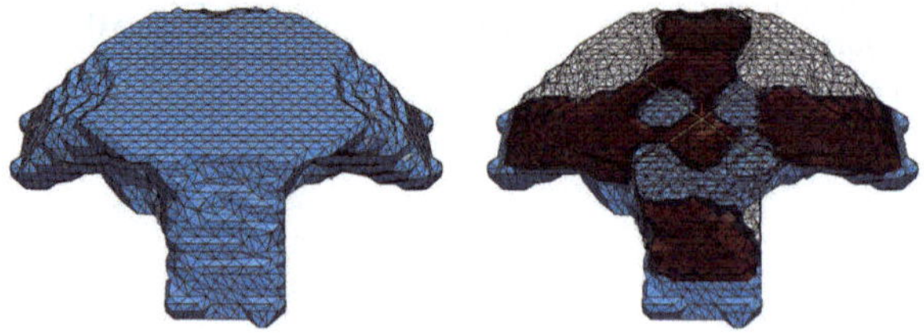

Fig. 5.67 Antenna configuration obtained using $w = 0.03$. *Top view* (*left image*) and *cut plane view* at 3.5 mm level (*right image*)

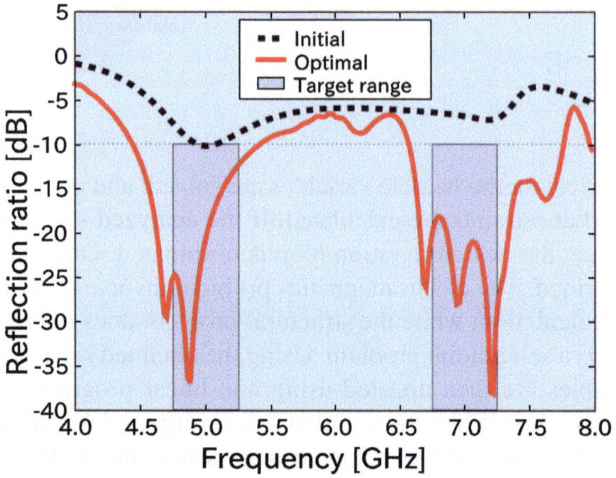

Fig. 5.68 Frequency characteristic of optimal antenna obtained using $w = 0.03$

frequencies. For reference, Fig. 5.69 also shows the antenna configurations with top, sectioned, and bottom views obtained for weighting values of 0, 0.5, 0.9, and 1 from left to right, respectively.

Fabrication Constraint

In the previous section, a multiband DRA without grayscale was successfully designed with a weighted sum multi-objective formulation and SIMP. However, the process for finding the appropriate weighting coefficient is inefficient. Additionally, as shown in Figs. 5.67 and 5.69, the obtained designs are hard to realize since they contain irregular cavities and holes. In this section, a problem formulation that provides for a fabrication constraint is explained.

The constraint method is another well-known approach for multi-objective problems. The constraint method transforms the multi-objective optimization problem into a single objective problem by converting objectives into constraints with the exception of the one single main objective. In our case, the electromagnetics objective is converted into a constraint since the maximum allowable reflection energy is easier to estimate than the maximum allowable elastic compliance. In this way, it is not necessary to calculate the Pareto front to find an optimum design that satisfies the

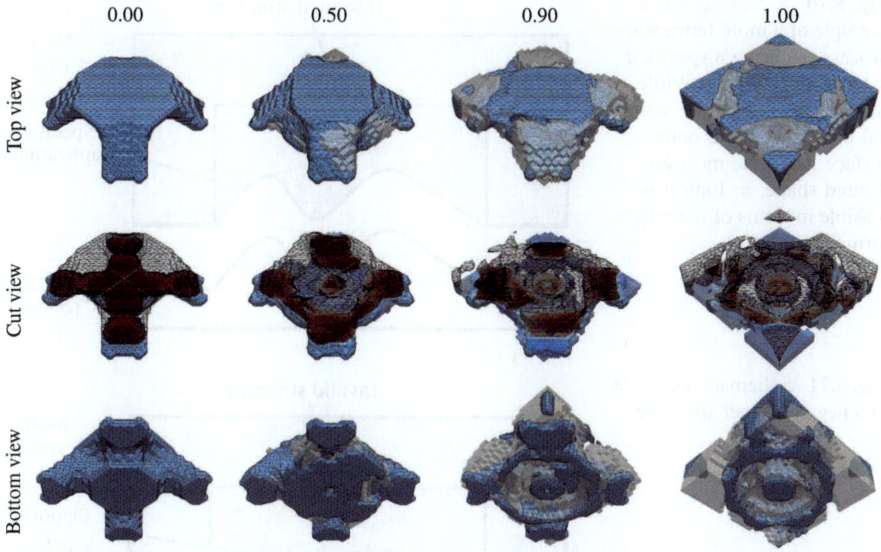

Fig. 5.69 Antenna configurations for $w = 0$, 0.5, 0.9, and 1 shown in columns from *left* to *right*, respectively. *Top view* (*top row*), *cut view* (*middle row*), and *bottom view* (*lower row*)

designer's requirement. Thus, the first term in Eq. (5.99b) is converted as follows:

$$\bar{\mathsf{F}}_o = \frac{\mathscr{P}_{ref}}{\mathscr{P}^{init}_{ref}} < \frac{\mathscr{P}_u}{\mathscr{P}^{init}}, \tag{5.102}$$

where $\mathscr{P}_u$ is the maximum allowable reflection energy.

For grayscale elimination, we introduce an exclusive volume constraint. The exclusive volume is the volume of the area that is occupied by any kind of material, including very thin density material. This is defined as follows:

$$v_{ex} = \int_D \mathsf{H}(\gamma(\mathbf{x}))\mathrm{d}\Omega, \tag{5.103}$$

where H is the Heaviside step function. When the exclusive area is completely filled with solid material, there exists no grayscale. This condition can be written as

$$v_{ex} = v, \tag{5.104}$$

where v is the volume ($\int_D \gamma \mathrm{d}\Omega$) commonly used in topology optimization, e.g. Eq. (5.99d). In the implementation, this expression is rewritten in a relaxed form

$$\tilde{v}_{ex} - v = \int_D \tilde{\mathsf{H}}(\gamma(\mathbf{x}))\mathrm{d}\Omega - \int_D \gamma \mathrm{d}\Omega < \varepsilon_{inf}, \tag{5.105}$$

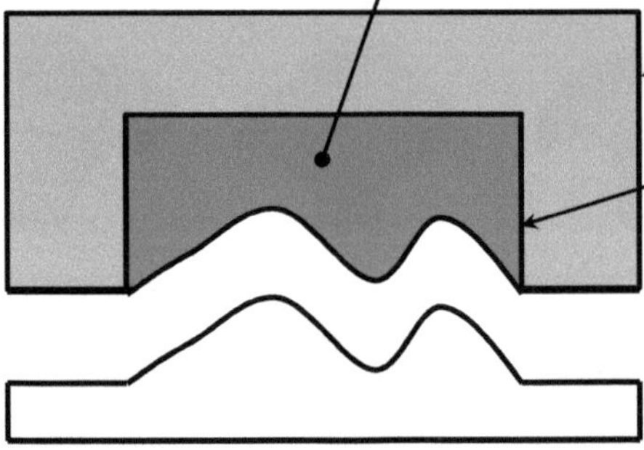

Fig. 5.70 A representative example of a mold fabrication process requiring a specified external appearance. While shown with a 'box-like' external appearance, the outer surface may take most any desired shape, as long it is feasible in terms of manufacturing

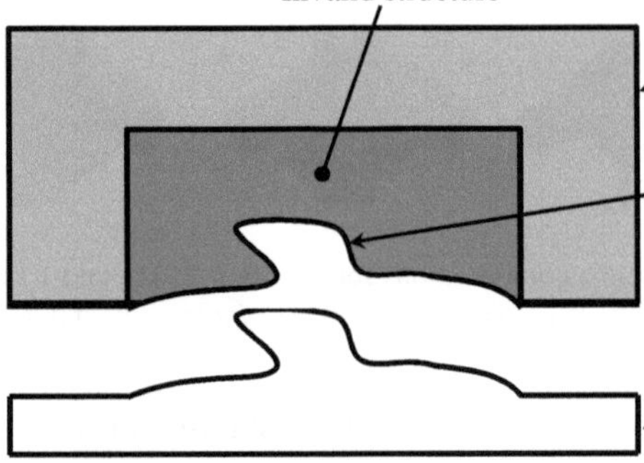

Fig. 5.71 Schematic example of a negative taper structure

where $\tilde{\mathsf{H}}$ is the relaxed Heaviside function, Eq. (4.9), introduced in Chap. 4, and ε_{inf} is an infinitesimal value. By enforcing the above constraint, grayscale can be eliminated from the design.

Assume that the antenna is made of high permittivity thermoplastic material, such as ceramic filled polymer composite, and is fabricated using molding technologies; refer to Fig. 5.70. The outer shape should reside in a specified area, due to cosmetic or packaging concerns. Also, the inner or bottom shape should have no negative taper to assure the product can be removed from the mold; see Fig. 5.71.

For the outer surface, one straightforward approach is to prepare a fixed design domain which satisfies the specification. However, this is not very flexible for adopting specification changes. Instead, the following constraints may be used:

$$\gamma(\mathbf{x}) < 0 \quad \text{if} \quad \bar{Q}(\mathbf{x}) < 0, \tag{5.106}$$

where, $\bar{Q}(\mathbf{x})$ is a level set function to represent a given outer shape ($\bar{Q}(\mathbf{x}) > 0$ is the domain side), thus $\bar{Q}(\mathbf{x})$ itself is constant throughout the optimization. Note that the design domain is discretized using a structured cubic grid since FDTD is used for the electromagnetics analysis, thus finding the adjacent design variable is trivial in contrast with using a finite element mesh.

Negative taper occurs if a point has more material density than a point above the point of interest. Therefore, in order to eliminate the negative taper, the following constraint may be used:

$$\gamma(\mathbf{x}-\Delta z)-\gamma(\mathbf{x})<0 \quad \text{if} \quad \bar{Q}(\mathbf{x}) \geq \Delta z, \tag{5.107}$$

where Δz is the grid size in z-direction. Note that the condition, $\bar{Q}(\mathbf{x}) \geq \mathbf{\Delta z}$ is a necessary and sufficient condition to avoid numerical contradiction, so that it can be replaced by other necessary conditions, as we show in the following complete formulation.

Here, we specify $\bar{Q}(\mathbf{x})$ as a spherical shell with center at $\mathbf{x}_c$ and radius r_0, that is $\bar{Q}(\mathbf{x}) = r_0 - |\mathbf{x}-\mathbf{x}_c|$. By combining all of the constraints above, the optimization problem may be formulated as follows:

$$\text{Find} \quad \gamma \tag{5.108a}$$

$$\underset{0<\gamma<1}{\text{Minimize}} \quad \mathsf{F}_o = \frac{l(\mathbf{u})}{l(\mathbf{u})^{init}} \tag{5.108b}$$

$$\text{Subject to} \quad \bar{\mathsf{F}}_o = \frac{\mathcal{P}_{ref}}{\mathcal{P}_{ref}^{init}} < \frac{\mathcal{P}_u}{\mathcal{P}^{init}}, \tag{5.108c}$$

$$v_{ex} = v \qquad \text{Exclusive volume} \tag{5.108d}$$

$$R_1 = \gamma(\mathbf{x}) < 0 \;\; \text{if} \;\; r_0 < |\mathbf{x}-\mathbf{x}_c| \qquad \text{Outer shape control (void domain)} \tag{5.108e}$$

$$T_1 = \gamma(\mathbf{x}-\Delta\mathbf{z}) - \gamma(\mathbf{x}) < 0 \;\; \text{if} \;\; |\mathbf{x}-\mathbf{x}_c| < r_0 - \Delta r \qquad \text{Taper control} \tag{5.108f}$$

Maxwell's equations; refer to Chap. 3, Sect. 3.6 (5.108g)

Eq. (5.97) (5.108h)

Given Eqs. (5.100)–(5.101). (5.108i)

Here, constraint Eqs. (5.108c) and (5.108d) are non-linear, whereas Eqs. (5.108e) and (5.108f) are linear constraints, where $\Delta r \geq \Delta z$. SLP is used as the non-linear programming method, the same as the previous problems. For the non-linear constraints, an augmented Lagrangian method is used. The linear constraints in the formulation are pointwise constraints so that the total number of them is approximately the same as the number of design variables. These constraints are enforced in subproblems in SLP.

Figure 5.72 shows the flowchart for the full optimization problem, Eq. (5.108). Compared with the previous flowchart, this process is more simple with only a single

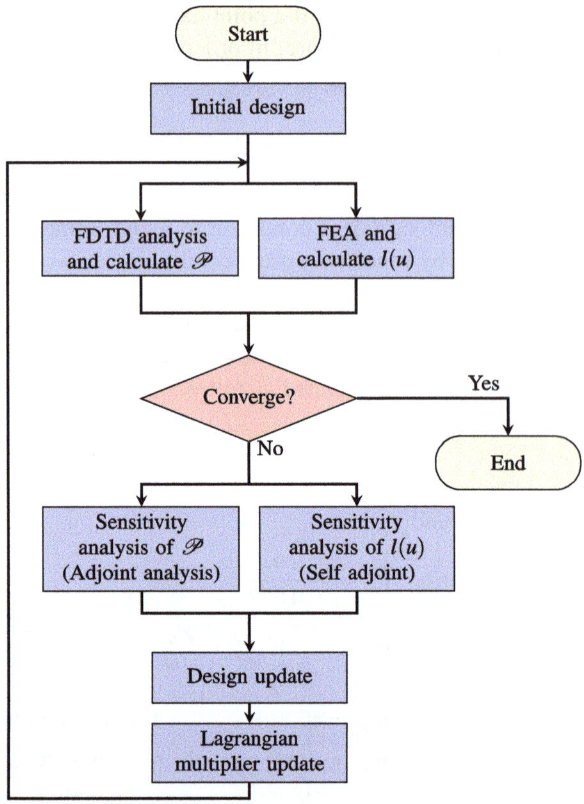

Fig. 5.72 Flowchart for a multiphysics antenna design optimization problem with fabrication constraints

loop, thus it is more computationally efficient. However, the number of the iterations using a single loop process may take more time because of the non-linear constraints. Still, this modified process is beneficial since the user does not have to decide an appropriate weighting coefficient set to cover the Pareto front.

Figure 5.73 shows the optimized configuration with and without the fabrication constraints. The case without the fabrication constraints is solved without Eqs. (5.108d), (5.108e) and (5.108f), but instead, with a normal upper volume constraint, i.e., Eq. (5.99d). The upper row shows a view from the top side, the middle row shows the same top side view with a cross section obtained by a vertical cut plane, and the lower row shows a view from the bottom side. The left column shows the result without fabrication constraints, while the right column shows the result with fabrication constraints. Both methods produce a configuration that has a small amount of grayscale area. This indicates the constraint-based approach for grayscale control is a better way to balance two objectives, in this case. The cross-sectional view reveals that the left side configuration has small cavities or inverse taper, whereas

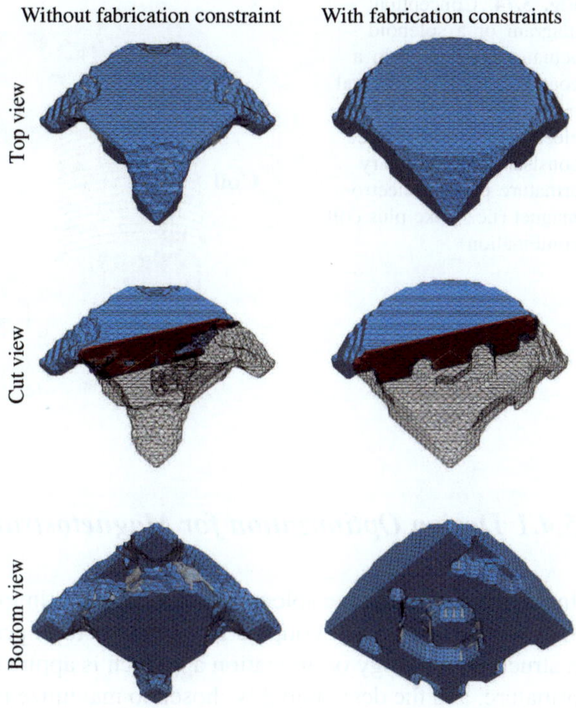

Fig. 5.73 Antenna configurations obtained with and without fabrication constraints

the right hand configuration does not, and maintains a mechanically stable structure with supports at the four bottom corners.

5.4 Actuator Analysis and Design

In this section, the analysis and design optimization of actuators is explained. Two separate 2-D numerical examples are provided covering common topics including the magnetostructural design of a solenoid and the design of an actuator with respect to the optimal arrangement of the permanent magnet, coils, and ferromagnetic material. The multiphysics nature of the first study elucidates the relationship between the magnetic force generated by a solenoid and the resulting structural deformation that occurs within the system. The second study then provides a detailed example of how to optimally arrange multiple components in an actuator to maximize the average magnetic force acting on the moving component (i.e., plunger). Here, the magnetic field is logically connected to the structural response. Both studies have clear ramifications in the development electromechanical assemblies, where overbuilt and bulky structures that utilize excessive amounts of expensive rare-earth materials are detrimental to the design of highly efficient systems.

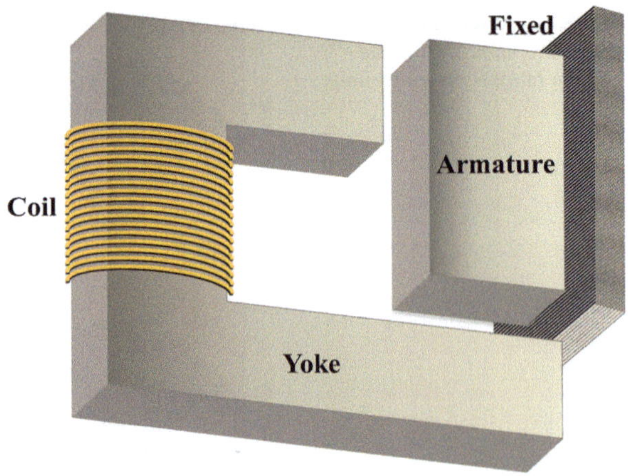

Fig. 5.74 Conceptual diagram of a solenoid actuator considered in a coupled magnetostructural problem. The electromechanical device consists of a stationary armature plus an electromagnet (i.e., yoke plus coil combination)

5.4.1 Design Optimization for Magnetostructural Coupling

In the first example, the solenoid actuator shown in Fig. 5.74 is optimally designed for the improvement of coupled magnetostructural performance. More specifically, a structural topology optimization approach is applied for the design of an actuator armature, and the design goal is chosen to maximize the total magnetic force acting on the armature and to minimize the mechanical deformation of the armature caused by the distributed magnetic force [56]. A magnetostatic analysis is performed first, and the distribution of the magnetic force is obtained. Next, the structural analysis is performed when the calculated force distribution is applied to the actuator. Finally, the mechanical compliance is computed. For the design optimization, the sensitivity analysis is performed using the adjoint method, and the optimization problem is formulated and solved using the sequential linear programming (SLP) method.

There are two distributed force calculation methods: (a) surface force distribution and (b) body force distribution. Using the first method, the force distribution on the surface of the magnetic materials is found, while using the second method, the distributed body force inside the magnetic materials is calculated using a virtual air-gap scheme [12, 13]. Both methods utilize general force calculation formulations such as Maxwell Stress Tensor (MST) or Coulomb Virtual Work (CVW) schemes. There has been inconclusive debate about whether the surface or body force distribution is a truer representation of the real situation. The surface and body force distributions obtained from the same magnetic field distribution are completely different. Thus, the mechanical deformation generated by two force distribution results may be different, as well. Here, the design optimization results for both surface and body force distributions are compared. Differences in optimization results are identified, and it is validated that each result is optimal in its force distribution. Such an example makes clear the multiphysics nature of actuator design.

Magnetostructural Analysis

The coupled magnetostructural analysis is performed using the finite element method. The analysis steps can be summarized as follows: (1) The magnetostatic analysis is carried out using the magnetic vector potential, **A**; (2) The magnetic force distribution, $\mathbf{f}_{M\_d}$, is calculated from the magnetic field distribution; (3) The magnetic force distribution yields the structural nodal force vector, $\mathbf{f}_S$. Then, the full structural analysis for the magnetic body force gives the mechanical displacement vector, **U**, and the compliance, l. The detailed procedure for each step is explained below.

For the first step, the magnetostatic governing equation (see Sect. 3.5) is formulated in matrix form using the finite element method as

$$\mathbf{K}_M \mathbf{A} = \mathbf{F}_M, \tag{5.109}$$

where $\mathbf{K}_M$ is the magnetic stiffness matrix, **A** is the nodal vector for magnetic vector potential, and $\mathbf{F}_M$ is the magnetic load vector. After solving Eq. (5.109), the magnetic flux density **B** is calculated as

$$\mathbf{B} = \nabla \times \mathbf{A}. \tag{5.110}$$

Second, from the magnetic field, **B**, the surface or body force distribution, $\mathbf{f}_b$, is calculated using either the MST or CVW method. In this example, the MST method is applied for computing both surface and body force distributions. For the calculation of the surface force distribution, the MST formulation is applied along the surface of the magnetic material. For the body force distribution, a virtual air-gap is inserted between the small bodies contacting each other because the MST method is defined only on an object surrounded by air. The magnetic flux density, **B**, is decomposed into normal and tangential components along the surface lines as follows:

$$\mathbf{B} = B_n \hat{\mathbf{n}} + B_t \hat{\mathbf{t}}, \tag{5.111}$$

where $\hat{\mathbf{n}}$ is a normal unit vector and $\hat{\mathbf{t}}$ is a tangential unit vector with respect to the surface lines. Then, the MST method gives the distributed magnetic force

$$\mathbf{f}_{M\_d} = \left[\oint \frac{1}{\mu_o} (B_n B_t) ds \right] \hat{\mathbf{n}} + \left[\oint \frac{1}{2\mu_o} (B_n^2 - B_t^2) ds \right] \hat{\mathbf{t}}. \tag{5.112}$$

Third, dividing this expression by the small body volume provides the distributed body force, $\mathbf{f}_b$, which is used to obtain the structural nodal force vector, $\mathbf{f}_S$. In the finite element formulation, the nodal force at the j-th node, $\mathbf{f}_S^j$, is represented as

$$\mathbf{f}_S = \int N_j \mathbf{f}_b d\Omega, \tag{5.113}$$

where N_j is the shape function of j-th node. Then, the finite element equation for the structural analysis can be written as

$$\mathbf{K}_S\mathbf{U} = \mathbf{F}_S, \tag{5.114}$$

where $\mathbf{K}_S$ is the assembled structural stiffness matrix, and $\mathbf{U}$ is the mechanical displacement vector. The assembled structural force vector, $\mathbf{F}_S$, is a function of the magnetic vector potential, $\mathbf{A}$. Therefore, the structural analysis is coupled with the magnetostatic analysis.

By solving Eq. (5.114), $\mathbf{U}$ is obtained, and finally the mechanical compliance, l, is calculated as

$$l = \mathbf{F}_S^{\mathrm{T}}\mathbf{U}. \tag{5.115}$$

When the compliance is minimized, the structural displacement is reduced, and consequently, the stiffness of the structure is globally maximized.

Structural Design Optimization

Structural topology optimization is applied in designing a solenoid armature. The design variables are the densities of each finite element in the design domain, and the material property is interpolated using the density method, as discussed in Chap. 4. The optimization goals are to minimize the mechanical compliance and to maximize the total magnetic force of the system at the same time. The total magnetic force, F_{tot}, is the magnitude of the total force vector which is obtained as the sum of the distributed body force. To achieve the above two design goals, a multi-term objective function for the optimization problem is formulated as a function of the design density variable, γ, as

$$\mathsf{F}_o = s_1[\mathbf{F}_S(\mathbf{A}(\gamma), \gamma)^{\mathrm{T}}\mathbf{U}(\gamma)] - s_2[F_{tot}(\mathbf{A}(\gamma), \gamma)], \tag{5.116}$$

where the objective function, F_o, is defined as the scaled sum of the mechanical compliance and the total magnetic force.

The scaling factors, s_1 and s_2, are controlled in each iteration step using the adaptive scaling strategy proposed in [86] and written as

$$s_1 = s_{1o}\frac{1/(F_{tot} + \varepsilon_{inf})}{1/(\mathbf{F}_S^{\mathrm{T}}\mathbf{U} + \varepsilon_{inf}) + 1/(F_{tot} + \varepsilon_{inf})}, \quad \text{and} \tag{5.117}$$

$$s_2 = s_{2o}\frac{1/(\mathbf{F}_S^{\mathrm{T}}\mathbf{U} + \varepsilon_{inf})}{1/(\mathbf{F}_S^{\mathrm{T}}\mathbf{U} + \varepsilon_{inf}) + 1/(F_{tot} + \varepsilon_{inf})}. \tag{5.118}$$

As opposed to the manual weighting of each objective function term, as performed in earlier sections of this chapter, this adaptive scaling strategy aims to avoid a local minimum satisfying only one objective. The scaling factors, s_1 and s_2, for each term of the objective function are, respectively, determined using the reciprocal value of the other term of the objective function. When the compliance, $\mathbf{F}_S^{\mathrm{T}}\mathbf{U}$, is dominantly reduced, the scaling factor, s_2, increases to reduce the total force, F_{tot}, and vice-versa. The term ε_{inf} is an additional infinitesimal number, which is utilized to avoid a 'divided-by-zero' situation. Then, the optimum solution that satisfies both objective function terms can be obtained.

In sum, the optimization problem for the coupled magnetostructural problem can be formulated as

Find	γ
Minimize	Eq. (5.116)
Subject to	Eqs. (5.109) – (5.115)
	$\int_{\Omega_d} \gamma \mathrm{d}\Omega_d - v_1 \le 0$
	$0 \le \gamma \le 1$
Given	$\mu_r = \mu_{r\_steel}\gamma^p$
	$E = E_{steel}\gamma^p$.

Here, the volume of the design domain is constrained to a specified volume fraction v_1, μ_r is the relative magnetic permeability, and E is the elastic modulus. The above optimization problem is solved by SLP, which requires the design sensitivity.

For this example, the sensitivity of the objective function is analytically derived using the adjoint variable method. The derivative of the function F_o with respect to the density design variable, γ, is written as

$$\frac{d\mathsf{F}_o}{d\gamma} = s_1\left[\frac{d(\mathbf{F}_S(\mathbf{A}(\gamma),\gamma)^{\mathrm{T}}\mathbf{U}(\gamma))}{d\gamma}\right] - s_2\left[\frac{d(F_{tot}(\mathbf{A}(\gamma),\gamma))}{d\gamma}\right]. \tag{5.119}$$

By using the adjoint variable method with two weakly coupled finite element equations, Eqs. (5.109) and (5.114), the first differentiation term in Eq. (5.119) is derived as

$$\frac{d\mathbf{F}_S^{\mathrm{T}}\mathbf{U}}{d\gamma} = 2\frac{\partial \mathbf{F}_S(\mathbf{A}(\gamma),\gamma)^{\mathrm{T}}}{\partial\gamma}\mathbf{U} + \hat{\lambda}_1^{\mathrm{T}}\frac{\partial \mathbf{K_M}}{\partial\gamma}\mathbf{A} + \hat{\lambda}_2^{\mathrm{T}}\Big(-\frac{\mathbf{K_S}}{\partial\gamma}\mathbf{U}\Big). \tag{5.120}$$

The adjoint variables, $\hat{\lambda}_1$ and $\hat{\lambda}_2$, are calculated by solving the following adjoint equations:

$$\mathbf{K}_M\hat{\lambda}_1 = -2\frac{\partial \mathbf{F}_S^{\mathrm{T}}}{\partial \mathbf{A}}\mathbf{U}, \quad \text{and} \tag{5.121}$$

$$\mathbf{K}_S\hat{\lambda}_2 = \mathbf{F}_S. \tag{5.122}$$

The second differentiation term of Eq. (5.119) is also derived using the adjoint variable method as

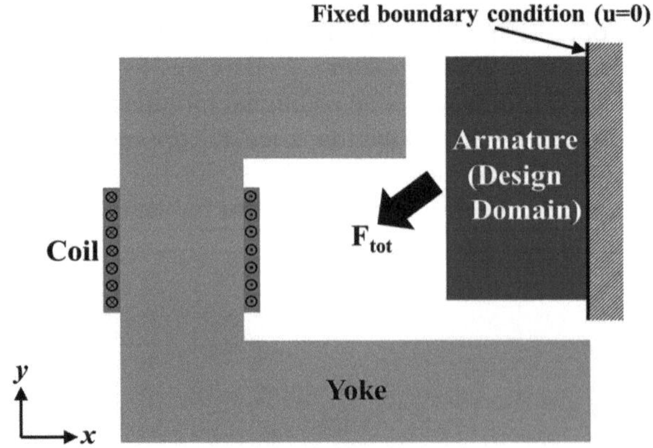

Fig. 5.75 Simplified 2-D magnetic actuator model and boundary condition for a magnetostructural solenoid design optimization problem.

$$\frac{dF_{tot}}{d\gamma} = \frac{\partial F_{tot}}{\partial \gamma} + \hat{\lambda}_3^{\mathrm{T}} \frac{\partial \mathbf{K}_M}{\partial \gamma} \mathbf{A}. \tag{5.123}$$

The adjoint equation necessary to calculate $\hat{\lambda}_3$ is then given as

$$\mathbf{K}_M \hat{\lambda}_3 = -\frac{\partial F_{tot}^T}{\partial \mathbf{A}}. \tag{5.124}$$

Finally, the analytical sensitivity of the objective function is obtained from Eqs. (5.119)–(5.124).

Optimization Model and Results

The solenoid actuator shown in Fig. 5.74 is designed using the above analysis and optimization approach. First, the actuator is simplified into a 2-D model, as shown in Fig. 5.75. The design domain is defined at the armature, and the displacement of the armature right boundary is fixed to zero. The design volume fraction is set as 0.6, and the SLP move limit is set as 0.004.

The two armature design results with equipotential lines are presented in Fig. 5.76. The numerical model was discretized using approximately 9,200 linear quadrilateral elements. One iteration for the optimization routine required about 5 s to solve on a quad-core workstation with a 3.4 GHz processor and 8 GB of RAM. The first design, AD1 (shown on the left), is obtained when the distributed magnetic force is calculated as a body force, and the second design, AD2 (shown on the right), is obtained when it is calculated as a surface force. Both design results, AD1 and AD2, have a similar general structure configuration. The structure connecting the left upper and bottom boundaries is designed to maximize the total force, F_{tot}. Then, most of the magnetic field flows through the armature, as is visualized through the equipotential lines. On the other hand, the structure in a right upper region is designed to minimize the mechanical deformation.

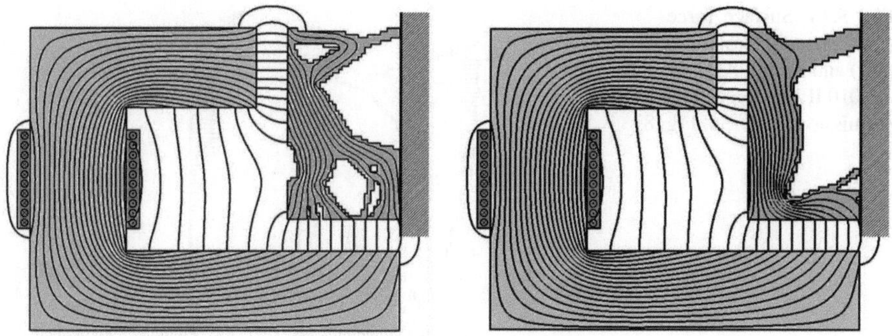

Fig. 5.76 Armature Design AD1 (*left*) obtained when the distributed magnetic force is calculated as a body force, and Armature Design AD2 (*right*) obtained when the distributed magnetic force is calculated as a surface force. Note that the magnetic equipotential lines are overlaid, for clarity. © 2010 IEEE. Reprinted with permission from [56, Figs. 5 and 6]

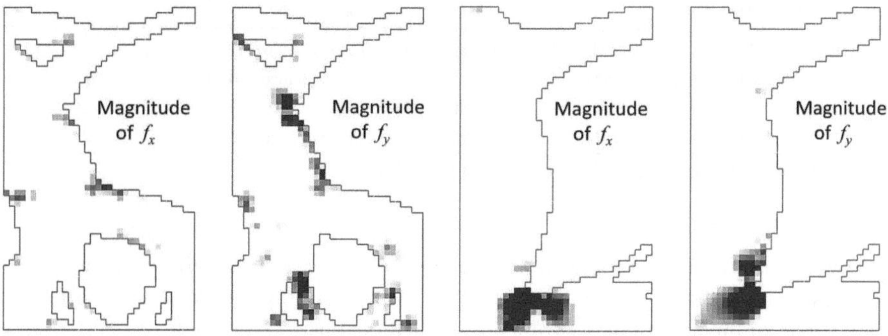

Fig. 5.77 Body force distribution of design AD1 (*left*) and design AD2 (*right*) in *grayscale*. © 2010 IEEE. Reprinted with permission from [56, Fig. 7]

Despite the similar general configuration, the two design results also show the following (considerable) differences. The design AD1 contains a large hole in the bottom part to raise the location connecting the armature and fixed boundary, while the design AD2 has a sharp corner in this same region. In addition, the design AD1 consists of a thin structure and small hole in the right upper region, while the design AD2 has a thick structure in this location. These differences are due to the minimization of the compliance via different force distributions.

To validate that each design result is optimal in its own force distribution, the body and surface force distribution of both design results are investigated. The body force distributions of the two design results AD1 and AD2 are presented as grayscale images in Fig. 5.77. The black colored regions correspond to a strong body force, and

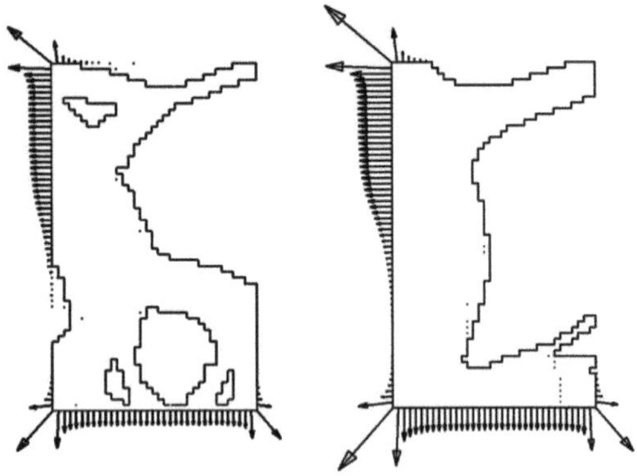

Fig. 5.78 Surface force distribution of design AD1 (*left*) and design AD2 (*right*). © 2010 IEEE. Reprinted, with permission from [56, Fig. 8]

Table 5.5 Comparison of the total magnetic force and mechanical compliance for AD1 and AD2 actuator designs

	Design AD1		Design AD2	
Body force	$F_{tot,x} = -2680.22$ N	$F_{tot,y} = -2680.22$ N	$F_{tot,x} = -2674.19$ N	$F_{tot,y} = -2047.13$ N
	Compliance = 0.6237×10^{-3} m/N		Compliance = 0.1355×10^{-2} m/N	
Surface force	$F_{tot,x} = -2690.49$ N	$F_{tot,y} = -2054.20$ N	$F_{tot,x} = -2685.01$ N	$F_{tot,y} = -2051.56$ N
	Compliance = 0.4747×10^{-3} m/N		Compliance = 0.2515×10^{-3} m/N	

© 2010 IEEE. Reprinted with permission from [56], Table I

the white color regions indicate a weak body force. The surface force distributions of two design results are presented in Fig. 5.78. Table 5.5 compares the x and y-direction total force and mechanical compliance of each design.

If the body force shown in Fig. 5.77 is the true force distribution, the design AD1 is much stiffer than the design AD2 as can be seen in the compliance comparison in Table 5.5. The body force of the design AD1 is evenly distributed around the center of the structure, and the compliance due to this distribution can be minimized by the raised connecting location. In contrast, a sharp corner of the design AD2 concentrates the body force, and consequently deteriorates the stiffness. On the other hand, if the surface force distribution shown in Fig. 5.78 is true, the design AD2 is much stiffer than the design AD1. The surface force of two designs is similarly distributed at the left upper and bottom boundaries. To minimize the compliance of this force distribution, a thick structure in the right upper part is required as can be seen in the design AD2. The total forces of all cases are almost identical as shown in Table 5.5. Therefore, both design AD1 and AD2 can satisfy the first design goal, that is, the maximization of the total force, whichever of the force distribution is true. However, each design can satisfy the compliance minimization only when the corresponding force distribution is true.

5.4.2 Simultaneous Design Optimization of Permanent Magnet, Coils, and Ferromagnetic Material

In this section, structural topology optimization is applied to the design of permanent magnet linear actuators, which are typically composed of PMs, coils, and ferromagnetic materials. The examples presented throughout this section deal with cases where multiple materials are simultaneously co-designed. The optimization of such multi-material systems is becoming increasingly important due to greater prevalence of both advanced composite materials as well as metamaterials in advanced electromechanical systems.

The optimization goal is to maximize the average magnetic force acting on a plunger as it travels over a distance of 20 mm. The magnetic force is calculated using the previously described MST method coupled with finite element analysis; refer to Sect. 3.5. The design sensitivity analysis is performed using the adjoint method, and the optimization problem is solved using an SLP method. The magnetic force calculation method and design optimization strategies are explained first. Then, two design results for PM linear actuators with different configurations are presented including: Design Result 1: The design of a moving actuator plunger composed of ferromagnetic material and PMs, and Design Result 2: The design of a stationary actuator yoke composed of PMs, ferromagnetic material, and positive plus negative direction coils for the situation where the actuator plunger is again moving.

Magnetic Force Calculation

The analysis step for the actuator magnetic force calculation is the same as described in the previous example in Sect. 5.4.1. The magnetic field distribution is first obtained using FEA, and the magnetic force is calculated using the MST method. For the magnetic field analysis, the magnetostatic governing equation governing equation (see Sect. 3.5) is formulated in a matrix form using the finite element method as

$$\mathbf{K}_M\mathbf{A} = \mathbf{F}_M, \tag{5.125}$$

where $\mathbf{K}_M$ is the magnetic stiffness matrix, $\mathbf{A}$ is the nodal vector for magnetic vector potential, and $\mathbf{F}_M$ is the magnetic load vector. After solving Eq. (5.125), the magnetic flux density $\mathbf{B}$ is calculated as

$$\mathbf{B} = \nabla \times \mathbf{A}. \tag{5.126}$$

The previous example in Sect. 5.4.1 deals with using the distributed force for the structural analysis. However, this example addresses only the total force acting on the plunger. Note that the integration of the distributed force along the body is the total magnetic force acting on the body. The MST method gives the total magnetic force, $\mathbf{F}_{tot}$, as

$$\mathbf{F}_{tot} = \left[\oint \frac{1}{2\mu_o}(B_n^2 - B_t^2)ds\right]\hat{\mathbf{n}} + \left[\oint \frac{1}{\mu_o}(B_n B_t)ds\right]\hat{\mathbf{t}}, \tag{5.127}$$

where B_n and B_t are, respectively, the normal and tangential components of magnetic flux density, $\mathbf{B}$, while $\hat{\mathbf{n}}$ and $\hat{\mathbf{t}}$ are, respectively, unit vectors normal and tangential to the integration path enveloping the body subject to the magnetic force, i.e., the actuator plunger.

The above magnetic force calculation process is repeated as the plunger is stepped through 20 equal 1 mm increments, which account for the full 20 mm plunger movement. In such a way, the average magnetic force acting on the plunger is computed over its entire range of travel.

Design Optimization

Here, the topology design optimization method is presented for the simultaneous co-design of the actuator PMs, coils, and ferromagnetic material. To apply topology optimization, the material properties in the design domain are interpolated using a method similar to that proposed in [14, 81]. Four design variables, γ_1 through γ_4, in each finite element control the material properties and allow for determination of the PM, coils or ferromagnetic material portions of the design domain. The relative permeability, μ_r, the residual magnetic flux density, $\mathbf{B}_r$, and the current density J_z are defined as

$$\mu_r(\gamma_1) = \mu_{r\_ferro}\gamma_1^{p_1} + \mu_{r\_air}(1 - \gamma_1^{p_1}) \tag{5.128}$$

$$\mathbf{B}_r = \gamma_2^{p_2}\mathbf{B}_{r\_PM} \tag{5.129}$$

$$J_z = \gamma_3^{p_3} J_z^* + \gamma_4^{p_3}(-J_z^*) \tag{5.130}$$

where p_1, p_2, and p_3 are the penalization parameters (respectively, set to 0.5, 3 and 3), μ_{r_ferro} and μ_{r_air} are the relative permeability of the ferromagnetic material and air (respectively, set to 1500 and 1), $\mathbf{B}_{r\_PM}$ is the residual flux density of the PM, and J_z^* is the loaded external current density at the coil. Here, the design variables γ_1 and γ_2 control the relative permeability and the strength of the PM of each finite element, respectively. Two design variables γ_3 and γ_4 are, respectively, used to represent equal volume positive and negative direction current density coils.

The design goal is to maximize the average axial direction magnetic force, $F_{tot\_x}^m$, acting on the plunger. Therefore, a objective function for the optimization problem can be expressed as

$$\mathsf{F}_o = \frac{1}{20}\sum_{m=1}^{20} F_{tot\_x}^m, \tag{5.131}$$

where the objective function, F_o, is defined as the average of the axial direction force, $F_{tot\_x}^m$, which can be calculated using Eq. (5.127) at 20 equally spaced plunger locations. The optimization problem is solved using the SLP method.

For this example, the sensitivity of the objective function, F_o, with respect to the design variables, γ_1 through γ_4, can be written as

$$\frac{d\mathsf{F}_o}{d\gamma_k} = \frac{1}{20}\sum_{m=1}^{20}\frac{dF_{tot\_x}^m}{d\gamma_k} \qquad (k = 1, ..., 4). \tag{5.132}$$

The sensitivity of the magnetic force at the m-th plunger location, $\frac{dF_{tot\_x}^m}{d\gamma_k}$, is analytically derived using the adjoint variable method. The magnetic force, $F_{tot\_x}^m$, with the adjoint term may be written as

$$F_{tot\_x}^m(\mathbf{A}^m) + \hat{\lambda}^{m^{\mathrm{T}}}[\mathbf{K}_M^m(\gamma_1)\mathbf{A}^m - \mathbf{F}_M^m(\gamma_1, \gamma_2, \gamma_3, \gamma_4)], \tag{5.133}$$

where $\hat{\lambda}^m$ is the adjoint variable. The differentiation of Eq. (5.133) with respect to γ_1 gives the sensitivity of F_{tot_x} as

$$\frac{dF_{tot\_x}^m}{d\gamma_1} = \hat{\lambda}^{m\mathrm{T}}\frac{\partial\mathbf{K}_M^m(\gamma_1)}{\partial\gamma_1}\mathbf{A}^m - \hat{\lambda}^{m\mathrm{T}}\frac{\partial\mathbf{F}_M^m(\gamma_1, \gamma_2, \gamma_3, \gamma_4)}{\partial\gamma_1}. \tag{5.134}$$

The adjoint variable, $\hat{\lambda}^m$, is obtained by solving the adjoint equation derived as

$$\mathbf{K}_M^{m\,\mathrm{T}}\hat{\lambda}^m = -\frac{\partial F_{tot\_x}^m(\mathbf{A}^m)^{\mathrm{T}}}{\partial\mathbf{A}^m}. \tag{5.135}$$

Likewise, the sensitivity of the objective function, F_o, with respect to γ_2-γ_4 is, respectively, derived as

$$\frac{dF_{tot\_x}^m}{d\gamma_k} = -\hat{\lambda}^{m\mathrm{T}}\frac{\partial\mathbf{F}_M^m(\gamma_1, \gamma_2, \gamma_3, \gamma_4)}{\partial\gamma_k} \qquad (k = 2, 3, 4). \tag{5.136}$$

The adjoint equation needed to calculate $\hat{\lambda}^m$ is derived following the prior procedure used to arrive at Eq. (5.135); refer to [58] for additional details.

Design Result 1: Linear Actuator Plunger Design

The above analysis and optimization method is applied to the design of the PM actuator model shown in Fig. 5.79. This actuator model is composed of a yoke with coils and a plunger that is composed of ferromagnetic material and PMs. The shape and location of the ferromagnetic material and PMs in the plunger is simultaneously optimized. In this example, the magnetization direction of the PMs is fixed in a vertical orientation. The material properties of the ferromagnetic material and PMs are interpolated using two design variables, γ_1 and γ_2; refer to Eqs. (5.128) and (5.129). The optimization goal is to maximize the average force during the 20 mm plunger movement. Thus, the optimization problem may be stated as

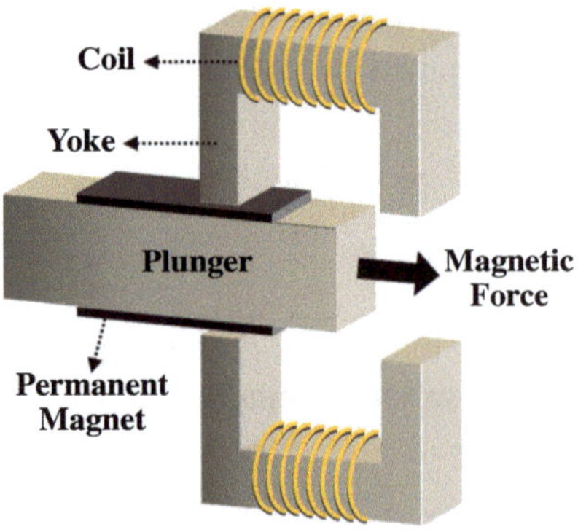

Fig. 5.79 Conceptual diagram of a linear actuator, Design 1, that consists of a stationary yoke with coil and a moving plunger with permanent magnets

Find	γ_1 and γ_2
Maximize	Eq. (5.131)
Subject to	Eqs. (5.125) – (5.127)
	$\int_{\Omega_d} \gamma_1 \mathrm{d}\Omega_d - v_1 \leq 0$
	$\int_{\Omega_d} \gamma_2 \mathrm{d}\Omega_d - v_2 \leq 0$
	$0 \leq \gamma_1 \leq 1$
	$0 \leq \gamma_2 \leq 1$
Given	Eqs. (5.128) – (5.129).

Here, the v_1 and v_2 are volume fraction for the ferromagnetic material and PM, respectively. For the optimization problem, the SLP move limit is set to 0.015. The volume constraints for the PMs and ferromagnetic material are, respectively, set to 5 % and 50 % of the design domain. An external current of 3 A flows through the coil, which is assumed to have 480 turns, and the residual magnetic flux density of the PM, B_{r_PM}, is set to 0.4 T.

The 2-D design and analysis domain for the PM actuator is shown in Fig. 5.80. The plunger movement is simulated by moving the plunger from $d = 0$ to $d = 20$ mm in 20 equal, 1 mm, steps. Due to symmetry, one-half of the actuator is analyzed using appropriate boundary conditions. In this example, PM actuators with two different configurations for the magnetic field source, i.e., the PM and external current are studied; see Fig. 5.81. The figure shows two combinations of the magnetization direction of the plunger PM and the direction of the current in the yoke coils. Model 1 has an anti-symmetric magnetic field source configuration, and thus one large magnetic field loop passes through all three components. For Model 1, a zero Neumann condition is applied to the vector potential, A_z, at the symmetry edge for zero tangential (i.e., x-direction) magnetic flux density, B_x. On the other hand, Model 2 has

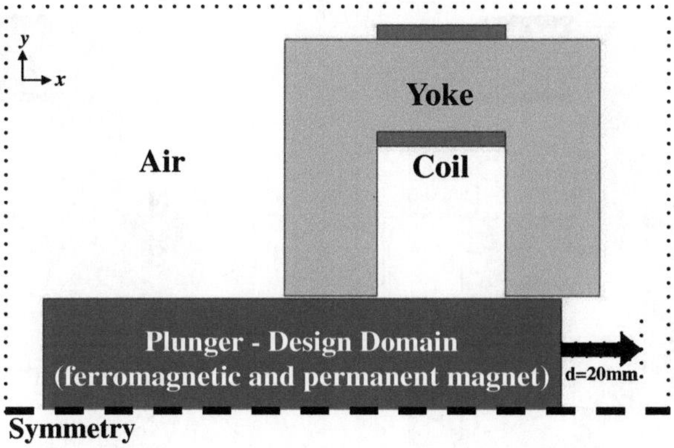

Fig. 5.80 Two-dimensional design and analysis domain for linear actuator Design 1

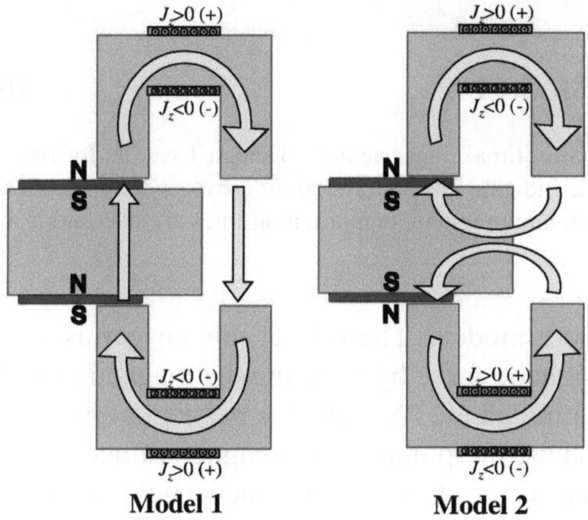

Fig. 5.81 Two linear actuator Design 1 models assumed to have different magnetic field source configurations (*initial designs shown*)

a symmetric field source configuration, which generates two symmetric and smaller magnetic field loops. For Model 2, a zero Dirichlet condition is applied to the vector potential, A_z, at the symmetry edge for a zero normal (i.e., y-direction) magnetic flux density, B_y. In both models, the optimal shape and locations of the PMs and ferromagnetic material in the plunger is found, and the average forces of the optimal designs are compared.

The plunger design results for the two models are compared in Fig. 5.82. In each case, the analysis model was discretized using approximately 9,000 linear quadrilateral elements, and a single iteration for the optimization loop required about 10 s to solve on a quad-core workstation with a 3.4 GHz processor and 16 GB of RAM. Observe in Fig. 5.82 that the optimal shape and location of the PMs (dark gray structures) are almost identical. Also, a small air-gap region appears to the right of

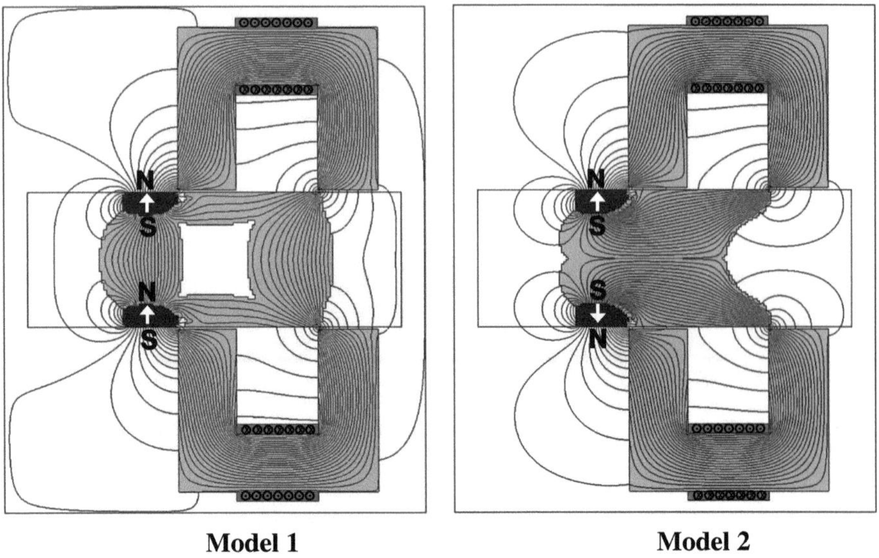

Fig. 5.82 Comparison of optimal linear actuator Design 1 results for two different models. Note that the *dark gray regions* indicate PM material, *light gray regions* indicate ferromagnetic material, *white regions* indicate air, and magnetic equipotential lines are overlaid, for clarity

the PM regions in both models. The role of this air-gap is to act as a flux barrier that increases the magnetic force by reducing the magnetic flux leakage that occurs inside the plunger. Although the PMs and flux barrier optimization results are almost identical in both models, the plunger ferromagnetic material (light gray structure) result is different due to the dissimilar magnetic field source configurations. The result for Model 1 shows a large cavity in the center of the plunger, but this cavity is not present in the results Model 2. The different optimization results, however, generate an almost identical magnetic flux distribution around the air-gap, as shown in Fig. 5.82. The magnetic field surrounding the air-gap mainly influences the magnetic force magnitude. Therefore, the different plunger designs result in a nearly identical optimal magnetic flux distribution around the air-gap, which results in a maximized magnetic force.

The force profiles with respect to the plunger locations are compared in Fig. 5.83. Table. 5.6 compares the average force from the initial design and optimized design. Although the average forces of two optimized designs (285.3 and 284.9 N) vary by less than 0.2 %, the initial design average forces (87.8 and 105.1 N) differ by 19.7 %. This demonstrates that the magnetic field source configuration affects the magnetic force in some structures, as expected. Lastly, in Model 1, the average force on the plunger is significantly increased by 325 %, while in Model 2 the average force is increased by 271 %.

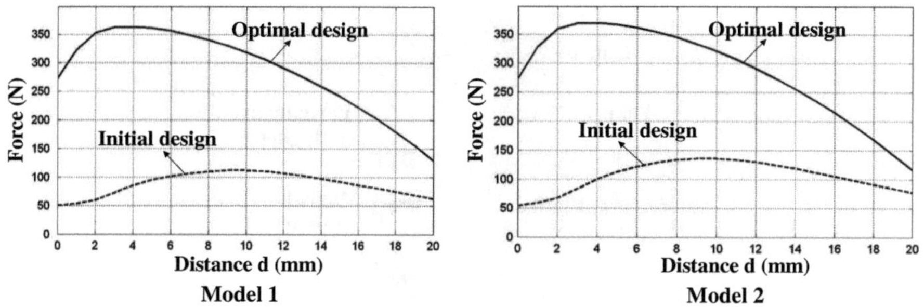

Fig. 5.83 Comparison of optimized and initial average force versus plunger distance traveled for linear actuator Design 1

Table 5.6 Average force comparison for linear actuator Design 1

	Initial design (Fig. 5.81)	Optimal design (Fig. 5.82)
Model 1	87.8 N	285.3 N
Model 2	105.1 N	284.9 N

Design Result 2: Linear Actuator Yoke Design

Following [58], the yoke of the linear actuator shown in Fig. 5.84 is now designed using the above analysis and optimization method. This actuator is again composed of a plunger and a yoke that consists of PMs, coils (negative and positive directions), and ferromagnetic material. The PMs, coil, and ferromagnetic material regions are designed using four optimization variables, γ_1–γ_4. The material properties of the ferromagnetic core, PM, and coils are interpolated using Eqs. (5.128)–(5.130). Once more, the optimization goal is to maximize the average magnetic force acting on the linear actuator plunger as it travels through a distance of 20 mm. An analysis and optimization approach similar to that used for linear actuator Design 1 is used. Thus, the optimization problem can be formulated as

$$
\begin{aligned}
\text{Find} \quad & \gamma_1 - \gamma_4 \\
\text{Maximize} \quad & \text{Eq. (5.131)} \\
\text{Subject to} \quad & \text{Eqs. (5.125)} - \text{(5.127)} \\
& \int_{\Omega_d} \gamma_1 \mathrm{d}\Omega_d - v_1 \le 0, \quad \int_{\Omega_d} \gamma_2 \mathrm{d}\Omega_d - v_2 \le 0 \\
& \int_{\Omega_d} \gamma_3 \mathrm{d}\Omega_d - v_3 \le 0, \quad \int_{\Omega_d} \gamma_4 \mathrm{d}\Omega_d - v_4 \le 0 \\
& 0 \le \gamma_1 \le 1, \quad 0 \le \gamma_2 \le 1 \\
& 0 \le \gamma_3 \le 1, \quad 0 \le \gamma_4 \le 1 \\
\text{Given} \quad & \text{Eqs. (5.128)} - \text{(5.130)}.
\end{aligned}
$$

Here, the volume constraints for the PM, positive direction coil, negative direction coil, and ferromagnetic material are set to 0.015, 0.025, 0.025, and 0.25, respectively.

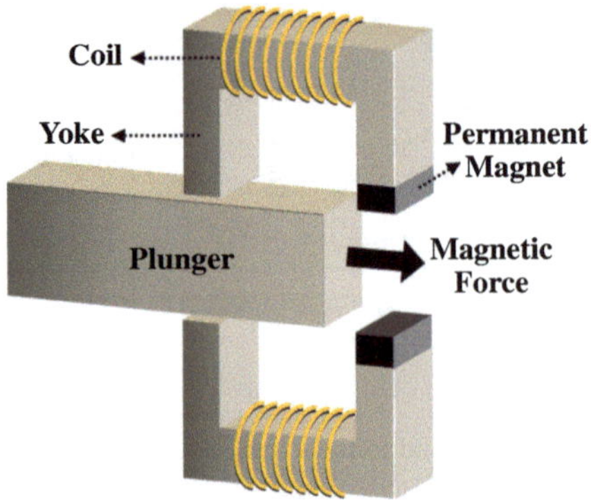

Fig. 5.84 Conceptual diagram of a linear actuator, Design 2, that consists of a stationary yoke with coils and PMs plus a moving plunger

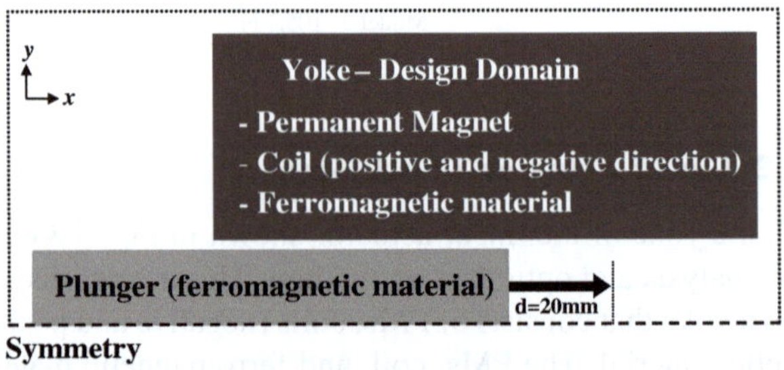

Fig. 5.85 Two-dimensional design domain and finite-element analysis domain with boundary conditions for linear actuator Design 2. © 2011 IEEE. Reprinted with permission from [58, Fig. 3]

In Eq. (5.130), an external current of 3 A flows through the coil, which is assumed to consist of 200 turns, and the residual magnetic flux density of the permanent magnet, B_{r_PM}, is similarly set to 0.4 T for Eq. (5.129).

The 2-D design and analysis domain of the Design 2 actuator is shown in Fig. 5.85. The plunger movement is simulated by moving the plunger from $d = 0$ to $d = 20$ mm in 20 equal, 1 mm, steps. Due to symmetry, one-half of the actuator is analyzed using appropriate boundary conditions. In this example, a zero Neumann condition is applied to the vector potential, A_z, at the symmetry edge for the zero tangential (i.e., x-direction) magnetic flux density, B_x. Using the topology optimization method, the optimal location and shape of the yoke components are found. In addition, the effect of the PM magnetization direction and the external current density strength on the optimization results is also investigated.

The initial and optimized results for linear actuator Design 2 are presented in Fig. 5.86 on the top left and right, respectively. In this example, the model was discretized using 9,600 linear quadrilateral elements, where one iteration of the optimization routine took slightly longer, 12 s, to solve on a quad-core workstation with a 3.4 GHz processor and 16 GB of RAM.

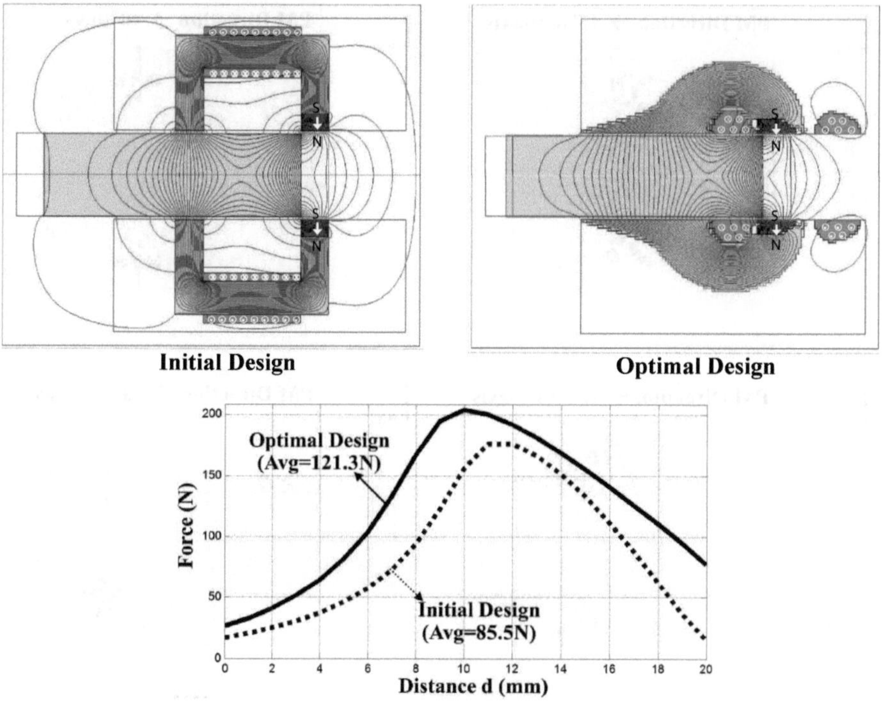

Fig. 5.86 Comparison of initial (*top left*) and optimal (*top right*) linear actuator Design 2 results at $d = 10$ mm, and their corresponding force profiles versus plunger location, d, (*bottom center*). Note that the *dark gray regions* indicate PM material, *light gray regions* indicate ferromagnetic material, *white regions* indicate air, coil regions are shown graphically, and magnetic equipotential lines are overlaid, for clarity. © 2011 IEEE. Reprinted with permission from [58, Fig. 4 and 5]

In the optimal structure in Fig. 5.86, both the PMs (dark regions) and the coils appear near the air-gap. The optimal structure maximizes the magnetic force by reducing the leakage magnetic field outside of the air-gap and plunger, which can be observed by comparing the magnetic flux lines of the initial and optimal designs in Fig. 5.86. Note that the magnetic field surrounding the air-gap mainly influences the magnetic force magnitude. The magnetic force profile with the plunger location, d, and the average force is also compared in Fig. 5.86 (bottom center image). The average force of the optimal design is expected to be 41.9 % higher than that of the initial design.

In addition, the influence of the PM magnetization direction on the optimization result is investigated here. The optimization process was repeated for four different magnetization directions (45°, 90°, 135°, and 180° with respect to the y-axis). Figure 5.87 shows the optimization results with magnetic flux lines for each magnetization direction. In every result, the coils appear near the air-gap to minimize leakage, and the PMs appear such that their magnetization directions are aligned with the direction of the magnetic field generated by the coil and passing through

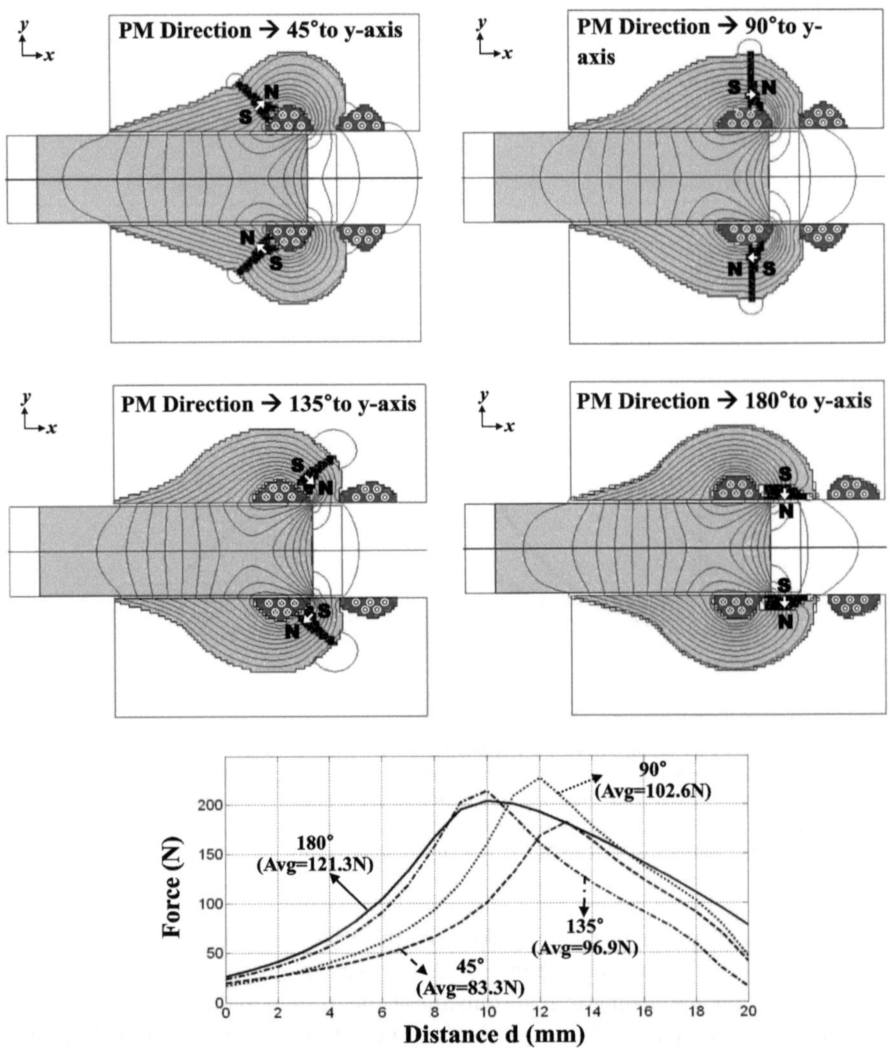

Fig. 5.87 Linear actuator Design 2 optimization results with magnetic flux lines (at d=10 mm) for different PM magnetization directions (*first and second rows*) and magnetic force profile comparison (*bottom center*). Note that the *dark gray regions* indicate PM material, *light gray regions* indicate ferromagnetic material, *white regions* indicate air, coil regions are shown graphically, and magnetic equipotential lines are overlaid, for clarity. © 2011 IEEE. Reprinted, with permission, from [58, Fig. 6]

the ferromagnetic material. The magnetic force profile and average force for each actuator design is further compared in Fig. 5.87 (bottom center image). When the magnetization direction is 180°, the PMs are located closest to the air-gap and consequently, the smallest leakage occurs along with the highest average force on the plunger.

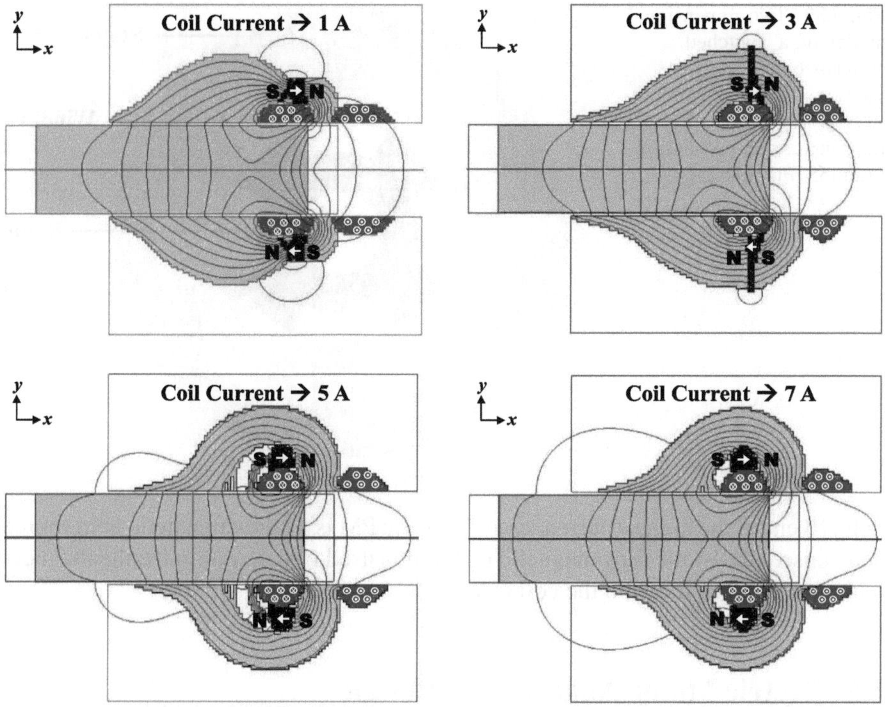

Fig. 5.88 Linear actuator Design 2 optimization results with magnetic flux lines (at d=10 mm) for a fixed PM strength and varying external current density strengths. Note that the *dark gray regions* indicate PM material, *light gray regions* indicate ferromagnetic material, *white regions* indicate air, coil regions are shown graphically, and magnetic equipotential lines are overlaid, for clarity. © 2011 IEEE. Reprinted with permission from [58, Fig. 7]

Finally, the effect of the external current strength on the optimization result is investigated. It is hypothesized that the ratio of the amount of PM to the external current density strength affects the optimization result. Thus, in order to investigate this effect, the optimization process is performed using four different current density strengths (i.e., 1 A, 3 A, 5 A, and 7 A) with a fixed PM strength of 0.4 T. Here, the magnetization direction is held fixed at 90° with respect to the y-axis. Figure 5.88 shows the various optimization results with magnetic flux lines. As the external coil current strength increases, the optimization process tends to generate a structure with smaller magnetic reluctance in order to maximize the magnetic field generated by the coils. Typically, a PM has high magnetic reluctance since its permeability is almost the same as that of air. Compared to the optimization result with 1 A, the structure with 3 A contains a thin PM structure, which has smaller magnetic reluctance despite higher leakage of the PM magnetic field. When the external current increases to 5 A, a thick ferromagnetic structure is formed to accommodate the magnetic field generated by the coil, and a thin additional structure is created for the PM magnetic

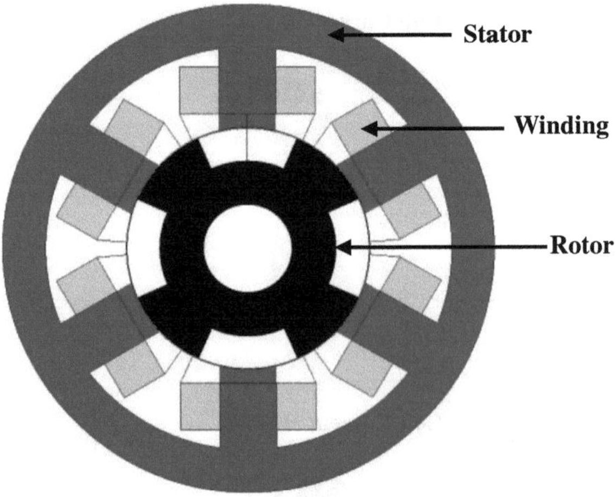

Fig. 5.89 Two-dimensional model of a switched reluctance motor with six stator poles and four rotor poles. © 2010 Lee J. Reprinted, with permission, from [55, Fig. 6.4.2a]

field. Even for the highest current case (7 A), the PM is a minor magnetic field source, and consequently, the ferromagnetic material is used only to minimize the magnetic reluctance associated with the coil source.

5.5 Electric Motor Analysis and Design

In this section, two different types of electric motors are considered including: (1) switched reluctance motors and (2) interior permanent magnet motors. Here, switched reluctance motors (SRMs) are first considered as an alternative to interior permanent magnet motors due to their advantage of not requiring expensive rare-earth materials in their construction. However, one well-known challenge with SRMs is the torque ripple associated with their operation, and the minimization of this torque ripple for an SRM is considered as a design objective in Sect. 5.5.1. In contrast, while scarcity and supply chain issues drives the cost of rare-earth materials up, an interior permanent magnet motor exhibits a significantly higher torque to motor volume ratio (or torque density) due to the enhanced magnetic properties of the associated permanent magnets [31]; such motors are found in many applications where increased performance in a compact package is required. Thus, for completeness, the multiphysics analysis of a more traditional interior permanent magnet motor is presented in Sect. 5.5.2.

5.5.1 Design Optimization of Switched Reluctance Motors

The following SRM design optimization study utilizes a 2-D model, as shown in Fig. 5.89 [55]. The design goal is to minimize the previous described torque ripple effect, which again is the main disadvantage of a switched reluctance motor. To achieve the design goal, not only the rotor/stator geometry of the motor but the electric variables (voltage on/off angles) are optimally designed. In addition, the copper loss is restrained by introducing a constraint on the root-mean-square (RMS) value of the phase current. The optimization problem is formulated for the above objective and constraint. The optimization problem is solved by using an SLP method, and then the optimal material distribution of the rotor/stator and the optimal voltage on–off angles can be found. The 2-D model of a 6/4 SRM (6 stator poles and 4 rotor poles) is chosen in the following design example. The analysis model takes the magnetic saturation effect into consideration, which means that the magnetic permeability is dependent on the magnetic flux density.

Performance Analysis

Motor performance is explicitly represented using a mathematical approximation method. This analytical representation of the performance is derived for the analytical sensitivity, which is required for gradient-based optimization. The performance analysis is conducted using the following procedure. Step 1: the flux linkage, λ_{link}, is approximated with respect to current, i, and rotor angle, θ_r, by using Fourier series expansions and piecewise quadratic polynomials. Step 2: the phase current curve is obtained by solving the voltage equation of an equivalent circuit. Step 3: the torque profile is determined using the global virtual work method.

Step 1—Flux Linkage Model: The flux linkage model represents the relation between the flux linkage, λ_{link}, the current of the stator coil, i, and the rotor position, θ_r. To calculate the flux linkage, the magnetic energy, W_{mag}, at discrete current values and rotor positions is calculated by solving the governing equations for low-frequency electromagnetics with magnetic saturation effect, which is explained in Sect. 3.5. This non-linear equation is solved using the finite element method with Newton–Raphson iteration. From the equation solution, the magnetic energy, W_{mag}, can be obtained as

$$W_{\text{mag}} = \int_{allspace} \left(\int_0^B \mathbf{H} \cdot d\mathbf{B} \right) dv, \tag{5.137}$$

where $\mathbf{H}$ is magnetic field intensity and $\mathbf{B}$ is magnetic flux density. This process is performed at various phase current values and rotor positions, and the magnetic energy, $(W_{\text{mag}})_{jk}$, at a discrete j-th phase current and k-th rotor position is calculated.

The next step is to find the flux linkage curves with respect to the phase current, i, axis at fixed rotor position, θ_r. From the approximation using piecewise quadratic polynomials, the flux linkage, $(\lambda_{\text{link}})_{km}$, at a fixed k-th rotor position and m-th piecewise interval can be written as

$$(\lambda_{\text{link}})_{km}(i) = C_{1km} i^2 + C_{2km} i + C_{3km}, \tag{5.138}$$

where C_{1km}–C_{3km} are the polynomial coefficients. The reason why piecewise quadratic polynomials are chosen is as follows. First, any piecewise function may reasonably represent both a line and a curve. The second-order polynomial gives the analytical solution of the voltage equation. Finally, the derivative of the flux linkage,

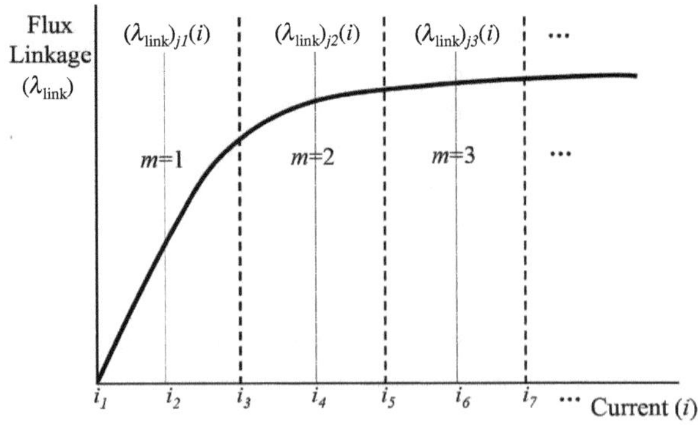

Fig. 5.90 Flux linkage, λ_{link}, curve with respect to the phase current, i, at a fixed rotor angle, θ_r, calculated using piecewise quadratic polynomials. © 2010 Lee J. Reprinted with permission from [55, Fig. 6.2.1]

λ_{link}, is not required for the performance analysis. Therefore, the C^1 continuity that can be considered in a cubic polynomial does not need to be satisfied. Figure 5.90 shows the flux linkage curve with respect to the phase current at a fixed rotor angle. To find the three coefficients, C_{1km}–C_{3km}, three conditions are required at each k-th rotor position and m-th piecewise interval. Two conditions are obtained from the relation between the flux linkage, $(\lambda_{link})_{km}$, and the magnetic energy, $(W_{mag})_{jk}$, which can be represented as

$$(W_{mag})_{jk} = i_j \cdot (\lambda_{link})_{km}\left[i_j\right] - \int_0^{i_j} (\lambda_{link})_{km} i\,di \tag{5.139}$$

In the above equation, the discrete current position, j, is set as $2m-1$, and $2m$. The last condition is the continuous condition at the current position, $j = 2m+1$. From these three conditions, the system of equations to find the coefficients C_{1jm}–C_{3jm} is derived as

$$\begin{bmatrix} \frac{2}{3}i_{2m}^3 + \frac{1}{3}i_{2m-1}^3 & \frac{1}{2}i_{2m}^2 + \frac{1}{2}i_{2m-1}^2 & i_{2m-1} \\ \frac{2}{3}i_{2m+1}^3 + \frac{1}{3}i_{2m-1}^3 & \frac{1}{2}i_{2m+1}^2 + \frac{1}{2}i_{2m-1}^2 & i_{2m-1} \\ i_{2m-1}^2 & i_{2m-1} & 1 \end{bmatrix} \times \begin{bmatrix} C_{1jm} \\ C_{2jm} \\ C_{3jm} \end{bmatrix} = \begin{bmatrix} A1 \\ A2 \\ A3 \end{bmatrix}, \tag{5.140}$$

where

$$\begin{aligned} A1 = {} & (W_{mag})_{j(2m)} \\ & + \sum_{k=1}^{m-1} (\frac{1}{3} C_{1jk}\left[i_{2k+1}^3 - i_{2k-1}^3\right] \\ & + \frac{1}{2} C_{2jk}\left[i_{2k+1}^2 - i_{2k-1}^2\right] + C_{2jk}\left[i_{2k+1} - i_{2k-1}\right]), \end{aligned} \tag{5.141}$$

$$A2 = (W_{mag})_{j(2m-1)}$$

$$+\sum_{k=1}^{m-1}(\frac{1}{3}C_{1jk}\left[i_{2k+1}^3-i_{2k-1}^3\right]$$

$$+\frac{1}{2}C_{2jk}\left[i_{2k+1}^2-i_{2k-1}^2\right]+C_{2jk}\left[i_{2k+1}-i_{2k-1}\right]), \quad (5.142)$$

$$A3 = C_{1j(m-1)}i_{2m-1}^2+C_{2jm}i_{2m-1}+C_{3jm}. \quad (5.143)$$

Next, a Fourier series expansion is applied to Eq. (5.138) for the approximation with respect to the rotor position, θ_r, axis. Then, the model representing the flux linkage, λ_{link}, with respect to phase current, i, and rotor angle, θ_r, is represented as

$$(\lambda_{\text{link}})_m(i,\theta_r) = \left(F_{im,0}i^2+F_{2m,0}i+F_{3m,0}\right)$$
$$+\sum_{n=1}^{NF}\left(F_{im,n}i^2+F_{2m,n}i+F_{3m,n}\right)\cos(nP_r\theta_r), \quad (5.144)$$

where NF is the number of Fourier series expansions, P_r is the number of rotor poles, and $F_{qm,n}$ are the coefficients of the Fourier series expansions, which can be written as

$$F_{qm,0}=\frac{1}{2\pi/P_r}\sum_{j=1}^{m}\left[(C_{q(j+1)m}+C_{q(j)m})(\theta_{r(j+1)}-\theta_{r(j)})\right], \quad (5.145)$$

$$F_{qm,n}=\frac{2}{n\pi}\sum_{j=1}^{m}\left[C_{q(j+1)m}\cdot\left(\sin(nP_r\theta_{r(j+1)})+\frac{\cos(nP_r\theta_{r(j+1)})-\cos(nP_r\theta_{r(j)})}{nP_r(\theta_{r(j+1)}-\theta_{r(j)})}\right)\right.$$
$$\left.-C_{q(j)m}\cdot\left(\sin(nP_r\theta_{r(j)})+\frac{\cos(nP_r\theta_{r(j+1)})-\cos(nP_r\theta_{r(j)})}{nP_r(\theta_{r(j+1)}-\theta_{r(j)})}\right)\right]. \quad (5.146)$$

Figure 5.91 shows the attained flux linkage model. The filled circular marks represent the flux linkage at discrete current and rotor angle positions, where the finite element analysis is carried out. As shown in Fig. 5.91, the flux linkage is smoothly approximated using the proposed method for the analytical representation with respect to current, i, and rotor angle, θ_r.

Step 2—Phase Current Curve: After calculating the flux linkage, the current curve with respect to rotor angle is obtained by solving the voltage equation. The voltage equation of SRMs is given by

$$V = Ri+\frac{d(\lambda_{\text{link}}(i,\theta_r)}{dt}, \quad (5.147)$$

where V is the source voltage, and R is the stator coil resistance.

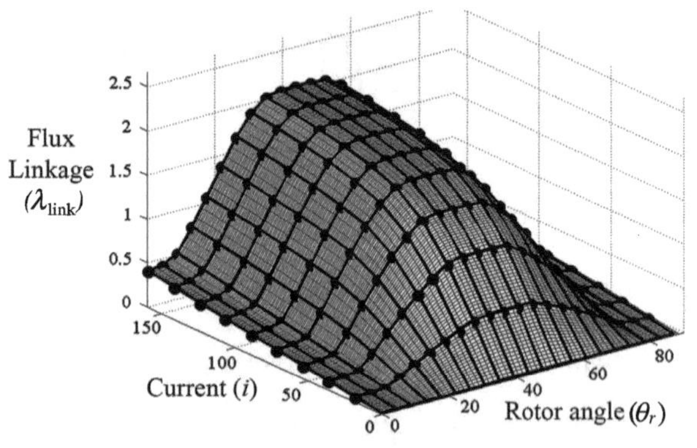

Fig. 5.91 Flux linkage, λ_{link}, model with respect to the phase current, i, and rotor angle, θ_r. © 2010 Lee J. Reprinted with permission from [55, Fig. 6.2.2]

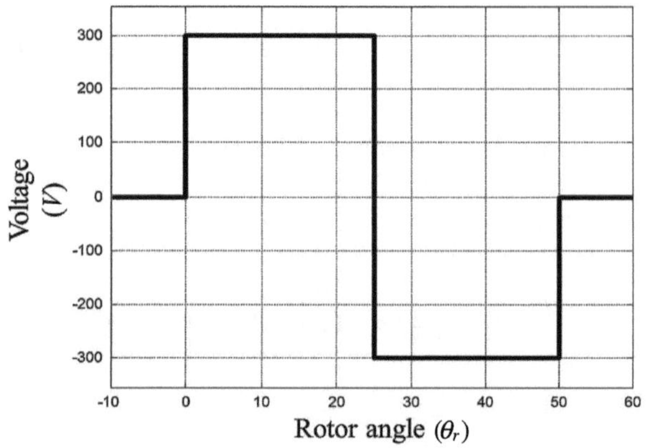

Fig. 5.92 Single pulse voltage waveform. The voltage on-off angles are set to 0° and 50 °, respectively. © 2010 Lee J. Reprinted, with permission, from [55, Fig. 6.2.3]

In order to solve the voltage equation analytically, the voltage drop due to the stator coil resistance is neglected. This assumption might cause errors in the performance analysis result. However, the voltage drop due to the resistance might be much less than the back electromotive force term by the time variation of the flux linkage. Thus, the error caused by the assumption is expected to be acceptable. Additionally, we assume steady-state rotation and zero phase current at the voltage-on angle. The voltage waveform used in this work is shown in Fig. 5.92. By solving Eq. (5.147) with the prior assumptions, the explicit expression of phase current curve is obtained as

$$i(\theta_r) = \frac{-X_2 \pm \sqrt{X_2^2 - 4X_1\left(X_3 - \lambda_{\text{link}(0)} - \frac{V_0}{\omega}(\theta_r - \theta_{r(0)}\right)}}{2X_1}, \tag{5.148}$$

where

$$X_i = \sum_{n=0}^{NF} F_{im,n} \cos(nP_r\theta_r), \tag{5.149}$$

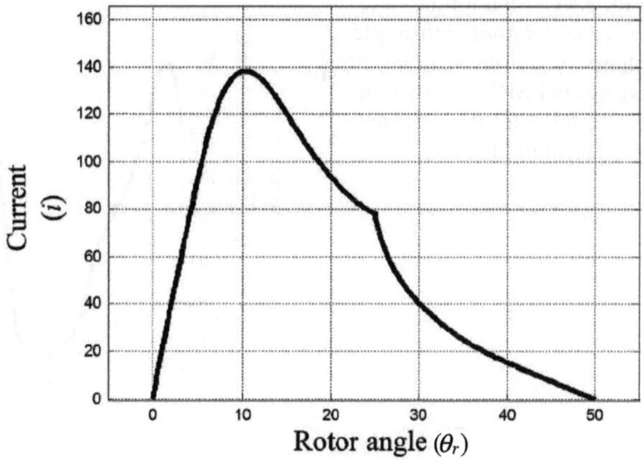

Fig. 5.93 Current curve of typical 6/4 SRM, Fig. 5.89, with voltage waveform in Fig. 5.92

$\theta_{r(0)}$ is the voltage-on angle, $\lambda_{\text{link}(0)}$ is the flux linkage at $\theta_{r(0)}$, and ω is the constant angular velocity [55]. Figure 5.93 shows the phase current profiles of typical 6/4 SRMs assuming non-linearly saturated material. In this example, typical rotor and stator shapes, as shown in Fig. 5.89, are used.

Step 3—Torque Curve: The torque profile of SRMs is calculated using the global virtual work method. The global virtual work method is based on the principle of conservation of energy and virtual displacement. This method uses the co-energy at a set of closely spaced rotor positions. The co-energy, W_{co}, is defined as

$$W_{co} = \int_0^i \lambda_{\text{link}} d\tilde{i} \bigg|_{\theta_r=\text{constant}}. \tag{5.150}$$

By taking the derivative of the total co-energy with respect to the rotor angle, θ_r, the torque, $\mathscr{T}$, profile is obtained as

$$\mathscr{T}(i, \theta_r) = \frac{\partial W_{co}}{\partial \theta_r}\bigg|_{i=\text{constant}}. \tag{5.151}$$

Substituting Eqs. (5.144) and (5.150) into Eq. (5.151), the torque profile is explicitly given by

$$\begin{aligned}
\mathscr{T}(i, \theta_r) = &-\sum_{n=0}^{NF}\Big[\frac{1}{3}F_{1m,n}(i^3 - i_{2m-1}^3) + \frac{1}{2}F_{2m,n}(i^2 - i_{2m-1}^2) \\
&+ F_{3m,n}(i - i_{2m-1})\Big] nP_r \sin(nP_r\theta_r) \\
&- \sum_{k=1}^{m-1}\sum_{n=0}^{NF}\Big[\frac{1}{3}F_{1k,n}(i_{2k+1}^3 - i_{2k-1}^3) + \frac{1}{2}F_{2k,n}(i_{2k+1}^2 - i_{2k-1}^2) \\
&+ F_{3k,n}(i_{2k+1} - i_{2k-1})\Big] nP_r \sin(nP_r\theta_r).
\end{aligned} \tag{5.152}$$

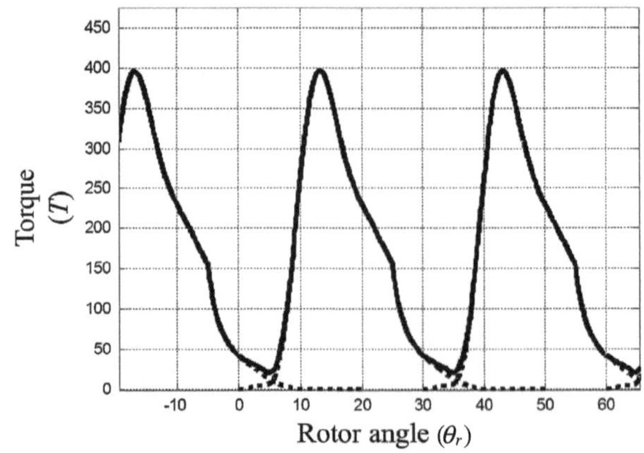

Fig. 5.94 Total torque, $\mathscr{T}_{tot}$, *curve* (*solid line*) with single-phase torque (*dashed line*) of typical 6/4 SRMs, shown in Fig. 5.89, with the voltage waveform in Fig. 5.92

It is noted that the function for the current, i, is obtained from Eq. (5.148). From Eq. (5.152), we can obtain the single-phase torque profile of SRMs (i.e., the torque profile when the voltage of a single pair of the stator poles is switched on and off). Since one phase is overlapped with the previous or the next phase in SRMs, the two consecutive single-phase torque profiles are summed in the overlapped range. After summing the torque profiles in all phases, the total torque profile is determined.

The overlapped range depends on the number of rotor poles, P_r, and stator poles, P_s. The rotor pole repeats its position after a rotor period of $2\pi/P_r$. During this period, the rotor pole passes $P_s/2$ stator poles. Therefore, the single-pulse torque profile is overlapped every P_{torque} calculated as

$$P_{torque} = \frac{2\pi/P_r}{P_s/2} = \frac{4\pi}{P_r P_s}. \tag{5.153}$$

In a steady rotation, the torque profiles of all phases are identical. Therefore, the total torque, $\mathscr{T}_{tot}$, is given by

$$\mathscr{T}_{tot}(\theta_r) = \sum_{m=1}^{2\pi/P_r} \mathscr{T}(\theta_r + (m-1)P_{torque}). \tag{5.154}$$

Figure 5.94 shows three consecutive single-phase torque profiles and the total torque profile of the SRM model shown in Fig. 5.89. In this analysis, the voltage on-off angles are set to 0° and 50°, respectively. Single-phase torque exists between these two angles, and repeats every 30° (i.e., P_{torque} is 30°). As illustrated in Fig. 5.94, SRMs have inherently large torque ripples.

Optimization Model and Results

As we have discussed throughout this book, structural topology optimization allows one to find the optimal arrangement of a structure by setting material densities as design variables. The material densities are relaxed as continuous variables in order to use a mathematical programming method such as SLP. The material becomes air when the density is zero, and it becomes steel when the density is one. The density distribution of a design domain is optimized to achieve specific design goals. Here, the design goals are to minimize torque ripple and the mass of the rotor

$$\mathsf{F}_o = \sum_{k=1}^{n} \left(\frac{\mathscr{T}_k}{\mathscr{T}^*} - 1 \right)^2 + w_1 \mathsf{F}_c(\gamma) + w_2 v(\gamma), \tag{5.155}$$

while the following constraint is applied to the RMS value of the current curve:

$$i_{rms} \left(\approx \sqrt{\frac{1}{\mathscr{T}} \sum_{k=1}^{n} i_k^2 dt} \right) \le i_{rms}^*, \tag{5.156}$$

where

$$\mathsf{F}_c(\gamma) = \sum_{u=1}^{ND} \frac{\gamma_u (1 - \gamma_u)}{ND}, \tag{5.157}$$

$$v(\gamma) = \frac{\sum_u \gamma_u A_u}{\sum_u A_u}, \tag{5.158}$$

$\mathscr{T}_k$ is the torque value at k-th rotor angle $\theta_{r(k)}$ ($k = 1, \ldots, n$), n is the number of discrete rotor angles, $\mathscr{T}^*$ is the target average torque, i_{rms} is the RMS value of phase current, A_u is the area of u-th element, w_1 and w_2 are the weighting values for the density convergence function, $\mathsf{F}_c(\gamma)$, and the normalized rotor mass function, $v(\gamma)$, respectively, and ND is the number of design variables.

To minimize the torque ripple, the torque, $\mathscr{T}_k$, at discrete rotor angles, $\theta_{r(k)}$, is handled to become the constant target average torque, $\mathscr{T}^*$. The density convergence function, $\mathsf{F}_c(\gamma)$, is added to the objective function to enforce the convergence of intermediate densities to zero or one, and the normalized rotor mass function, $v(\gamma)$, is added to minimize the mass of the rotor as an additional objective function. In the constraints, the RMS value of current curve is restricted to i_{rms}^* to confine the copper loss, which is one of main heat sources, i.e., ($\mathscr{P}_{copper} = R i_{rms}^2$). Thus, the optimization problem can be stated as

Find	γ_u
Minimize	Eq. (5.155) with Eqs. (5.157) – (5.158)
Subject to	Eq. (5.156)
	$\int_{\Omega_d} \gamma_u \mathrm{d}\Omega_d - v \le 0$
	$0 \le \gamma_u \le 1.$

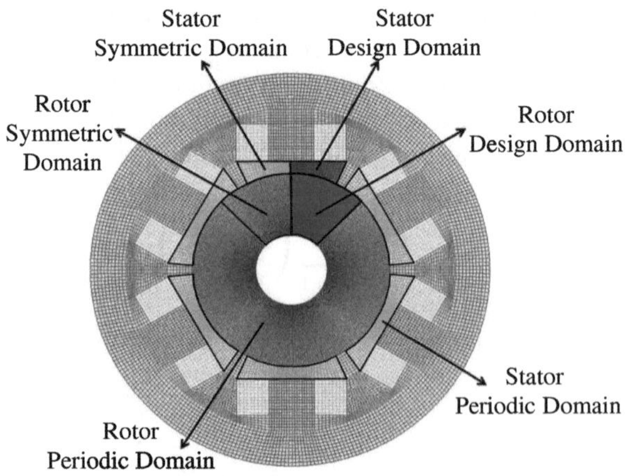

Fig. 5.95 Geometric design domain in rotor and stator (with symmetric/periodic domains plus finite element mesh for a 6/4 SRM). With kind permission from [57, Fig. 5], © Springer-Verlag 2010

The structural design domain is shown in Fig. 5.95 [57]. Since a 6/4 SRM is chosen, a 1/12 part of the stator and a 1/8 part of the rotor are considered as the geometric design domain. Then, the material densities in the geometric design domain are reflected and copied to the corresponding symmetric and periodic parts of the rotor and stator. Using this approach, the symmetries in the rotor and stator of the considered SRM are effectively exploited.

The formulated optimization problem is solved using the SLP method requiring the sensitivity of objective and constraint functions. The proposed performance analysis model gives the analytical representation of the current curve and torque profile. Therefore, the analytical sensitivity of objective and constraint functions with respect to design variables, γ_u, is available.

The sensitivity is analytically derived using direct differentiation and the chain rule. The sensitivity of the objective function, F_o, in Eq. (5.155) with respect to the design variable, γ_u, is given as

$$\frac{d\mathsf{F}_o}{d\gamma_u} = \sum_{j=1}^{n} 2\left(\frac{\mathscr{T}_k}{\mathscr{T}^*} - 1.0\right)\frac{\partial \mathscr{T}_k}{\partial \gamma_u} + w_1\frac{1-2\gamma_u}{ND} + w_2 A_u. \tag{5.159}$$

A 2-D 6/4 SRM is designed using the proposed optimization method. The target average torque, $\mathscr{T}^*$, is set as the average torque of the typical design, which is shown in Fig. 5.89. The constraint value, i^*_{rms}, on the RMS value of phase current is also set as the RMS current of the typical design. The move limits of the SLP method are 0.004 for the density design variables and 0.0004 for the angle design variables. The grayscale structure obtained as the optimization result is presented in Fig. 5.96. The dark colored regions represent steel, and the light colored regions represent air. The 2-D SRM computational model was discretized using 20,672 linear quadrilateral elements, and a single iteration of the optimization routine required roughly 9 s when solving the problem on a quad-core workstation with a 3.4 GHz processor and 8 GB of RAM.

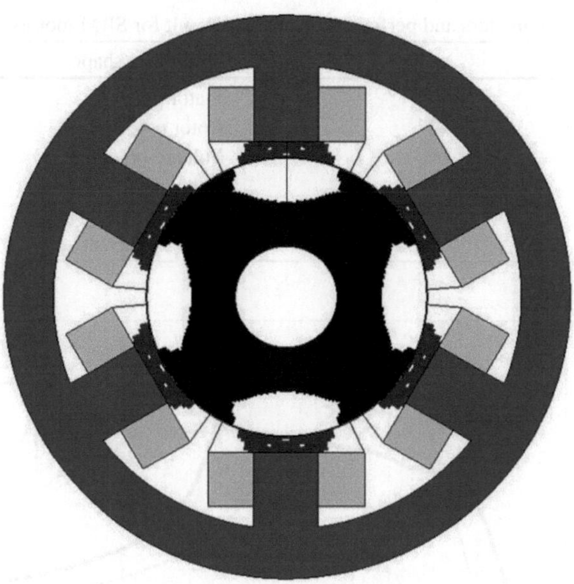

Fig. 5.96 Optimal topology of a 6/4 SRM rotor and stator. Note that *dark colored regions* indicate steel, *while light colored regions* indicate air; refer to Fig. 5.89 for further descriptions. © 2010 Lee J. Reprinted, with permission, from [55, Fig. 6.4.2c]

As shown in Fig. 5.96, the density design variables have clearly converged to zero (void) or one (steel) by applying the density convergence function, $\mathsf{F}_c(\gamma)$, in Eq. (5.157). The rotor and stator poles are optimized to have a notched shape near the air-gap. The notched shape of rotor and stator poles manipulates the inductance profile, and consequently, minimizes the SRM torque ripples.

Table 5.7 shows the design parameters of the typical and optimized SRM and also compares the average torques and the RMS values of the current curve. As shown in Table 5.7, the arc lengths of both the rotor and the stator poles are increased in order to minimize torque ripples. The range of the voltage on–off angles is also widened. The average torque of the optimized motor successfully satisfies the target average torque (e.g. 6 % lower than target average torque). The RMS value of the current curve in the designed motor is well constrained.

In Fig. 5.97, the current curve and total torque profile of the optimized design (solid lines) are compared to those of the typical design (dashed lines). As shown in Fig. 5.97a, the operating ranges of phase current increase with a widened voltage waveform. The narrow current curve of the typical model gets spread in the optimized model while satisfying the current constraint. As shown in Fig. 5.97b, the total torque profile of the optimized design is nearly matched with the target torque, and the torque ripple is remarkably reduced.

Table 5.7 Design parameters and performance analysis result for SRM motors

Typical shape		Optimized shape	
Stator pole arc	27°	Stator pole arc	39°
Rotor pole arc	42°	Rotor pole arc	45°
Voltage-on angle	0°	Voltage-on angle	−4.43°
Voltage-off angle	50°	Voltage-off angle	63.50°
Average torque	184.7 N m	Average torque	172.9 N m
RMS of current	57.2 A	RMS of current	56.4 A

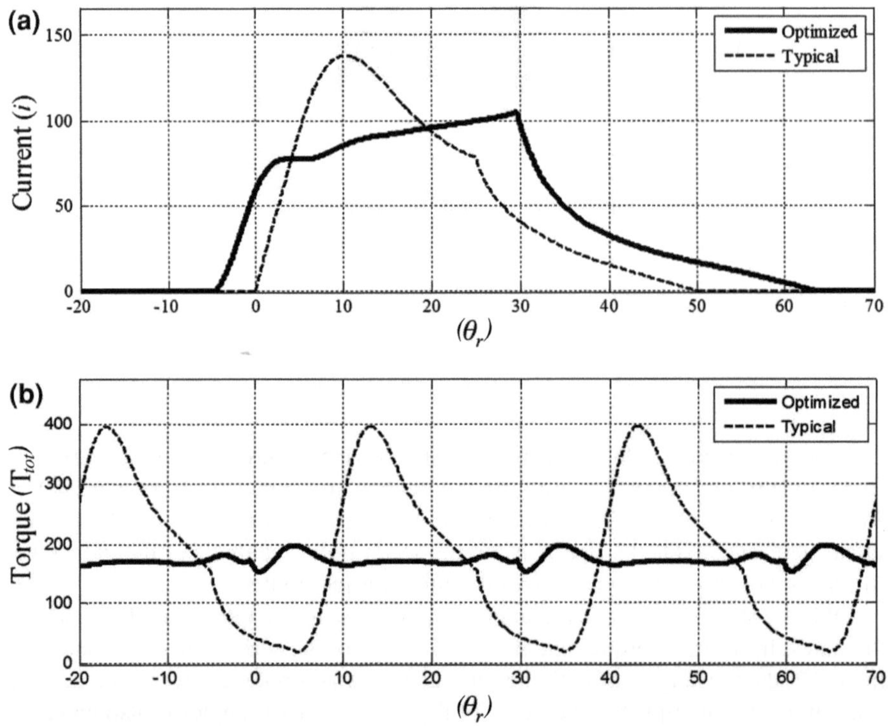

Fig. 5.97 Performance comparison of the typical 6/4 SRM shown in Fig. 5.89 and optimized 6/4 SRM shown in Fig. 5.96: **a** Current curve, **b** total torque profile. © 2010 Lee J. Reprinted with permission from [55, Fig. 6.4.3a, b]

5.5.2 Multiphysics Analysis of Interior Permanent Magnet Motors

The interior permanent magnet synchronous motor (IPMSM) has been the most promising candidate for the powertrain of hybrid and electric vehicles due its high power density and efficiency. In this final simulation example of the text, the multiphysics analysis of an IPMSM is presented. The geometry and material properties of the IPMSM used in this study is chosen based on the IPMSM found in a typi-

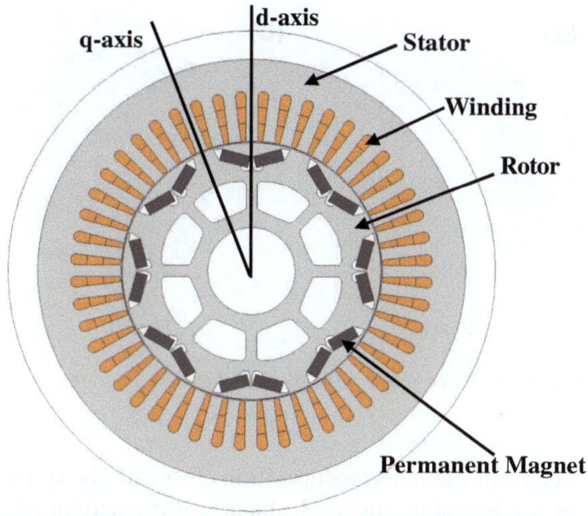

Fig. 5.98 Two-dimensional model of an interior permanent magnet synchronous motor (IMPSM) with 48 stator poles and 8 rotor poles

Table 5.8 Geometric parameters of the assumed IPMSM

Stator outer diameter	264 mm	Stator Inner diameter	161.9 mm
Rotor outer diameter	160.4 mm	Rotor Inner diameter	27.2 mm
Magnet length	17.88 mm	Magnet width	7.16 mm
Air-gap	0.73 mm	Lamination thickness	0.305 mm
Stack length (Thickness)	50.8 mm		

cal commercial hybrid vehicle [32, 78]. Here, the modeling of the IPMSM is first explained, and then a coupled electromagnetic/heat transfer analysis is performed to predict the torque and thermal performance of the IPMSM.

Modeling of IPMSM

In order to carry out the multiphysics simulation of an IPMSM, an analysis model was prepared using the commercial software COMSOL®[5] [17]. The geometric parameters and material properties of the IPMSM components are chosen based on the data presented in [32, 78]. Figure 5.98 shows the assumed 2-D geometry of the IPMSM. The motor consists of the 3-phase distributed windings, a 48-pole stator, and an 8-pole rotor with 16 NdFeB PM pieces. The axis in the middle of the rotor PM poles is called the direct-axis, or simply d-axis, and the intermediate axis between d-axes is called the quadrature axis, or simply q-axis. The q-axis is leading the d-axis by 90° electrical degrees.

The assumed geometric parameters of the IPMSM model are summarized in Table 5.8. The configuration of the three-phase winding is presented in Fig. 5.99. The phase A winding is located at the outermost position, and the phase C winding is

[5] COMSOL® is a registered trademark of COMSOL AB.

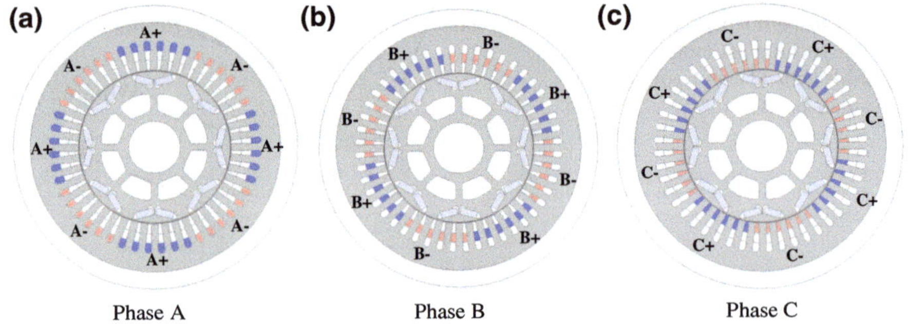

Fig. 5.99 Three-phase winding configuration diagrams

located at the innermost position closest to the rotor. As mentioned above, there are 48 winding slots, and each slot area is divided into three for a three-phase distributed winding. The winding for each phase is excited by an AC current with different phase angles. The current for each phase is modeled as

$$I_{\pm A} = \pm I_{peak} \cdot \sin(2\pi f t), \tag{5.160}$$

$$I_{\pm B} = \pm I_{peak} \cdot \sin(2\pi f t - 120°), \tag{5.161}$$

$$I_{\pm C} = \pm I_{peak} \cdot \sin(2\pi f t - 240°), \tag{5.162}$$

where I_{peak} is the amplitude of the AC current, and f is the frequency of the AC current, which is synchronous with the rotating speed of the rotor. When applying the above currents to each phase, the eight poles of the stator electromagnet rotates proportionally to the frequency of the AC current. The wire size is set as American Wire Gauge (AWG) 20, which has a diameter of 0.812 mm. The winding area for one phase in one slot is approximately 50 mm^2. In this example, the winding is modeled as 12 parallel bundled wires, and 3 bundled wires are positioned inside the area designated for one phase in one slot. When applying this wire configuration, the RMS current density is calculated as 10.2 A/m^2 when the phase current is 100 RMS Ampere.

The material physical parameters for the PM, iron, and coils are summarized in Table 5.9. Due to the magnetic saturation effect of the iron material, the relative permeability, μ_r, of the iron material gradually decreases as the magnetic flux density, B, increases. For the electromagnetic analysis, a quasistatic assumption is adopted for simplicity, and thus electrical conductivity, ς, and permittivity, ϵ, material physical parameters are not used in this example. For the thermal analysis, only steady-state heat conduction in solid materials is considered, and thus, the density, ρ, and heat capacity, C_p, material physical parameters are also not used.

Table 5.9 Material physical parameters for the various components in the assumed IPMSM

Material	Relative permeability (μ_r)	Thermal conductivity (k)
Iron in stator and rotor	1200 (when B is less than 1 T)	20 W/(m K)
PM in rotor	1 (Constant)	5 W/(m K)
Copper in winding	1 (Constant)	400 W/(m K)

Multiphysics Analysis

A coupled electromagnetic and thermal analysis is performed using the aforementioned commercial FEA software [17]. The 2-D IPMSM model was discretized using 75,658 second order triangular elements. Here, the single computational analysis required for the non-linear magnetostatic-thermal problem required approximately 60 seconds to solve on a quad-core workstation with a 3.4 GHz processor and 8 GB of RAM.

In the solution process, a low-frequency electromagnetic field analysis is first performed by solving the magnetostatic governing equations; refer to Sect. 3.5. A synchronous motor rotates by maintaining the electrical degrees between rotor poles and stator poles. Figure 5.100a, b allow for a comparison of the magnetic field distribution when the electrical degrees between the rotor PM pole and stator pole are 0° and 90°, respectively. As can be seen in Fig. 5.100, the electrical degrees are 0° when the quadrature axis (q-axis) current is a maximum and the direct axis (d-axis) current is zero, and they are 90° when the q-axis current is zero and the d-axis current is a maximum.

From the magnetic field distribution, the torque applied to the rotor is calculated using the Maxwell Stress Tensor method. Figure 5.101 presents the torque curves for the motor with respect to electrical degrees. The total torque of an IPMSM motor is composed of two components, which include the reluctance torque and PM torque. Reluctance torque is generated due to the interaction between the stator electromagnet and rotor iron. The magnetic reluctance of the rotor varies in different rotor positions because of the uneven shape of the rotor iron, which generates the torque. As shown in Fig. 5.101, the reluctance torque is zero when the stator current is matched with either the q-axis (i.e., electrical degrees = 0°) or d-axis (i.e., electrical degrees = 90°). PM torque is generated due to the interaction between the stator electromagnet and the rotor PM. PM torque is maximized when the stator current along the q-axis is maximized (i.e., electrical degrees = 0°), and it becomes zero when the stator current is aligned along the d-axis (i.e., electrical degrees = 90°). Because the periods of the torque components are different, the electrical angle maximizing the total torque is located between 0° and 90° depending on the amplitude of each torque term.

Next, the thermal analysis of the IPMSM is performed, where heat is generated by the coil loss in the winding plus core loss in the stator iron. In the steady-state operation of a synchronous motor, the magnetic field in the rotor winding does not vary with respect to time, and thus the core loss in the rotor iron is ignored in this example. The coil loss, $\mathscr{P}_{coil}$, in the winding can be calculated as Joule loss

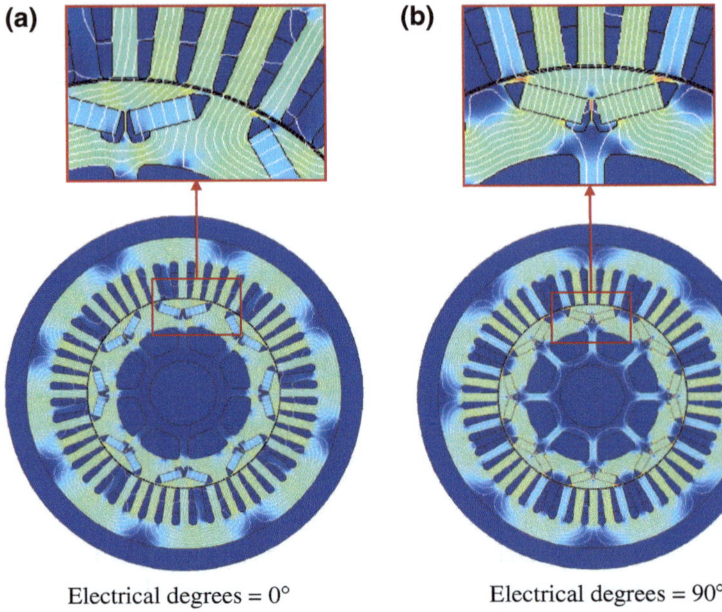

Fig. 5.100 Two-dimensional low-frequency electromagnetic field analysis results for an IPMSM with 48 stator poles and 8 rotor poles. Note that *lighter colored contours* indicate a higher current density, and the magnetic equipotential lines are overlaid, for clarity

Fig. 5.101 Torque curves for an IPMSM with respect to electrical angles

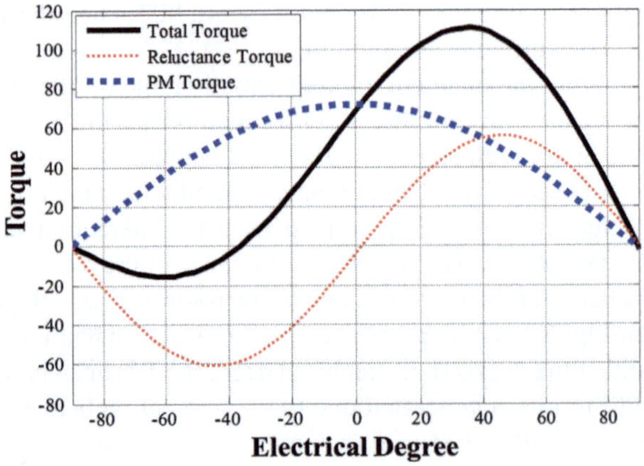

$$\mathscr{P}_{\text{coil}} = \int \rho_e J^2 dv, \tag{5.163}$$

where ρ_e is the electrical resistivity of the copper (assumed to be 1.68×10^{-8} Ohm·m) and J is the current density (in units of A/m^2) of the winding. The core loss in the stator iron can be divided into eddy current loss and hysteresis loss. The eddy current

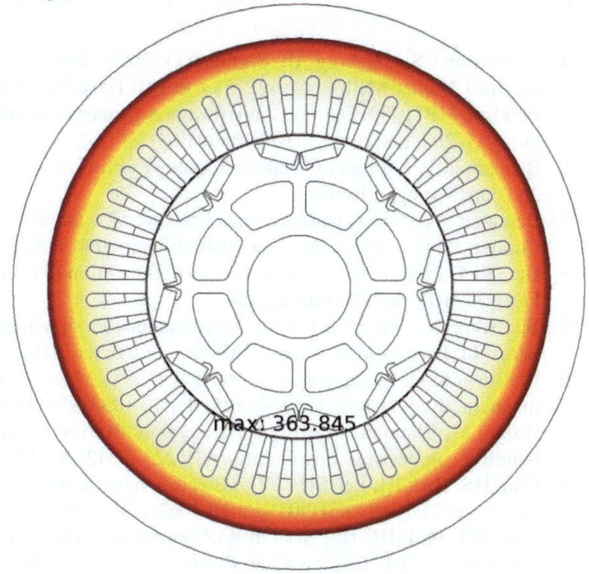

Fig. 5.102 Steady-state temperature distribution for an IPMSM based on off core loss results computed from the prior low-frequency electromagnetic analysis. Note that *lighter colored contours* indicate higher relative temperatures and the maximum temperature occurs at the end of the stator pole near the air-gap

and hysteresis losses are, respectively, calculated following [66] as

$$\mathscr{P}_{\text{eddy}} = \varsigma \frac{d^2}{12}\left(\frac{dB}{dt}\right), \tag{5.164}$$

$$\mathscr{P}_{\text{hysteresis}} = k_h B^2 f, \tag{5.165}$$

where ς is the electrical conductivity of the iron (1×10^7 S/m), d is the thickness of the laminated iron (0.305 mm), k_h is the hysteresis coefficient (set as 143 Ws/T^2/m^3), and f is the fundamental frequency (266.7 Hz corresponding to 2000 RPM). A convective cooling condition is applied at the motor outer boundary, where the heat transfer coefficient is set as 50 W/(m^2 K) with a reference temperature of 293.2 K. The distribution of the temperature is shown in Fig. 5.102, and the maximum temperature, 363.8 K, is expected at the end of stator pole near the air-gap.

References

1. Aage N, Mortensen NA, Sigmund O (2010) Topology optimization of metallic devices for microwave applications. Int J Numer Meth Eng 83:228–248. doi:10.1002/nme.2837
2. Alexandersen J, Andreasen CS, Aage N, Lazarov BS, Sigmund O (2013) Topology optimisation for coupled convection problems. 10th world Congress on structural and multidisciplinary optimization, Orlando, 19–24 May 2013
3. Andkjær J, Nishiwaki S, Nomura T, Sigmund O (2010) Topology optimization of grating couplers for the efficient excitation of surface plasmons. J Opt Soc Am B 27:1828–1832. doi:10.1364/JOSAB.27.001828

4. Andreasen CS, Gersborg AR, Sigmund O (2008) Topology optimization for microfluidic mixers. Int J Numer Meth Fl 61:498–513. doi:10.1002/fld.1964
5. Bendsøe MP (1989) Optimal shape design as a material distribution problem. Struct Multidiscip O 1:193–202. doi:10.1007/BF01650949
6. Bendsøe MP, Sigmund O (2003) Topology optimization: theory, methods, and applications, 2nd edn. Springer, Berlin
7. Berenger J (1994) A perfectly matched layer for the absorption of electromagnetic waves. J Comput Phys 114:185–200. doi:10.1006/jcph.1994.1159
8. Borvall T, Petersson J (2003) Topology optimization of fluids in stokes flow. Int J Numer Meth Fl 41:77–107. doi:10.1002/fld.426
9. Boulware JC, Jensen S (2010) Thermomagnetic siphoning on a bundle of current-carrying wires. In: Proceedings of the COMSOL conference 2010, Boston, 7–9 Oct 2010
10. Braithwaite D, Beaugnon E, Tournier R (1991) Magnetically controlled convection in paramagnetic fluid. Nature 354:134–136. doi:10.1038/354134a0
11. Bruggi M, Cinquini C (2011) Topology optimization for thermal insulation: an application to building engineering. Eng Optimiz 43:1223–1242. doi:10.1080/0305215X.2010.550284
12. Choi HS, Park IH, Lee SH (2006) Electromagnetic body force calculation based on virtual air gap. J Appl Phys 99:08H903. doi:10.1063/1.2173207
13. Choi HS, Park IH, Lee SH (2006) Concept of virtual air gap and its applications for force calculation. IEEE T Magn 42:663–666. doi:10.1109/TMAG.2006.871594
14. Choi JS, Yoo J (2009) Simultaneous structural topology optimization of electromagnetic sources and ferromagnetic materials. Comput Method Appl M 198:2111–2121. doi:10.1016/j.cma.2009.02.015
15. COMSOL AB (2008) COMSOL Multiphysics, Ver. 3.5a. Stockholm
16. COMSOL AB (2010) COMSOL Multiphysics, Ver. 4.0a. Stockholm
17. COMSOL AB (2011) COMSOL Multiphysics, Ver. 4.2. Stockholm
18. Cooper M, Petosa A, Wight JS, Ittipiboon A (1996) Investigation of dielectric resonator antennas for L-band communications. Paper presented at the antenna technology and applied electromagnetics symposium, ANTEM '96
19. Dede EM (2009) Multiphysics topology optimization of heat transfer and fluid flow systems. In: Proceedings of the COMSOL conference 2009, Boston, 8–10 Oct 2009
20. Dede EM (2010) Multiphysics optimization, synthesis, and application of jet impingement target surfaces. In: Proceedings of the 12th IEEE intersociety conference on thermal and thermomechanical phenomena in electronic systems, Las Vegas, 2–5 June 2010. doi:10.1109/ITHERM.2010.5501408
21. Dede EM (2010) The influence of channel aspect ratio on performance of optimized thermal-fluid structures. In: Proceedings of the COMSOL conference 2010, Boston, 7–9 Oct 2010
22. Dede EM (2010) Simulation and optimization of heat flow via anisotropic material thermal conductivity. Comp Mater Sci 50:510–515. doi:10.1016/j.commatsci.2010.09.012
23. Dede EM (2011) Experimental investigation of the thermal performance of a manifold hierarchical microchannel cold plate. In: Proceedings of the ASME 2011 Pacific Rim technical conference and exhibition on packaging and integration of electronic and photonic systems, MEMS and NEMS, vol 2. Portland, 6–8 July 2011. doi:10.1115/IPACK2011-52023
24. Dede EM (2012) Optimization and design of a multipass branching microchannel heat sink for electronics cooling. J Electron Packaging 134:041001. doi:10.1115/1.4007159
25. Dede EM (2012) Jet impingement heat exchanger apparatuses and power electronics modules. US Patent, 8,199,505 B2
26. Dede EM (2013) Power electronics modules and power electronics module assemblies. US Patent, 8,391,008 B2
27. Dede EM, Liu Y (2013) Experimental and numerical investigation of a multi-pass branching microchannel heat sink. Appl Therm Eng 55:51–60. doi:10.1016/j.applthermaleng.2013.02.038
28. Dede EM, Liu Y (2013) Cold plate assemblies and power electronics modules. US Patent, 8,427,832 B2

29. Dede EM, Nomura T, Schmalenberg P, Lee JS (2013) Heat flux cloaking, focusing, and reversal in ultra-thin composites considering conduction-convection effects. Appl Phys Lett 103:063501. doi:10.1063/1.4816775
30. Dede EM, Nomura T, Lee J (2014) Thermal-composite design optimization for heat flux shielding, focusing, and reversal. Struct Multidiscip O 49:59–68. doi:10.1007/s00158-013-0963-0
31. Dent PC (2012) Rare earth elements and permanent magnets (invited). J Appl Phys 111:07A721. doi:10.1063/1.3676616
32. Dorrell DG, Knight AM, Evans L, Popescu M (2012) Analysis and design techniques applied to hybrid vehicle drive machines—assessment of alternative ipm and induction motor topologies. IEEE T Ind Electron 59:3690–3699. doi:10.1109/TIE.2011.2165460
33. Erentok A, Sigmund O (2008) Three-dimensional topology optimized electrically-small conformal antenna. Paper presented at the IEEE antennas and propagation society international symposium, San Diego, 5–12 July 2008. doi:10.1109/APS.2008.4619200
34. Eshelby JD (1957) The determination of the elastic field of an ellipsoidal inclusion, and related problems. P R Soc A 241:376–396. doi:10.1098/rspa.1957.0133
35. Fan Z, Antar Y, Ittipiboon A, Petosa A (1996) Parasitic coplanar three-element dielectric resonator antenna subarray. Electron Lett 32:789–790. doi:10.1049/el:19960514
36. Finlaysan BA (1970) Convective instability of ferromagnetic fluids. J Fluid Mech 40:753–767. doi:10.1017/S0022112070000423
37. Garimella SV, Singhal V (2004) Single-phase flow and heat transport and pumping considerations in microchannel heat sinks. Heat Transfer Eng 25:15–25. doi:10.1080/01457630490248241
38. Gelet J-L, Gerlaud A (2013) Control of Joule heating extends performance and device life. IEEE Spectrum 5:S-24–25
39. Halbach K (1980) Design of permanent multipole magnets with oriented rare earth cobalt material. Nucl Instrum Methods 169:1–10. doi:10.1016/0029-554X(80)90094-4
40. Harpole GM, Eninger JE (1991) Micro-channel heat exchanger optimization. In: Proceedings of the 7th IEEE SEMI-THERM symposium, Phoenix, 12–14 February 1991. doi:10.1109/STHERM.1991.152913
41. Hasselman DPH, Bhatt H, Donaldson KY, Thomas JR (1992) Effect of fiber orientation and sample geometry on the effective thermal conductivity of a uniaxial carbon fiber reinforced glass matrix composite. J Compos Mater 26:2278–2288. doi:10.1177/002199839202601506
42. Hatta H, Minoru T (1986) Equivalent inclusion method for steady state conduction in composites. Int J Eng Sci 24:1159–1172. doi:10.1016/0020-7225(86)90011-X
43. Hull D, Clyne TW (1996) An introduction to composite materials, 2nd edn. Cambridge University Press, Cambridge
44. Iga A, Nishiwaki S, Izui K, Yoshimura M (2009) Topology optimization for thermal conductors considering design-dependent effects, including heat conduction and convection. Int J Heat Mass Tran 52:2721–2732. doi:10.1016/j.ijheatmasstransfer.2008.12.013
45. Incropera FP (1999) Liquid cooling of electronic devices by single-phase convection. Wiley-Interscience, New York
46. Incropera FP, DeWitt DP, Bergman TL, Lavine AS (2007) Introduction to heat transfer, 5th edn. Wiley, Hoboken
47. Jensen JS, Sigmund O (2004) Systematic design of photonic crystal structures using topology optimization: low-loss waveguide bends. Appl Phys Lett 84:2022. doi:10.1063/1.1688450
48. Jin N, Rahmat-Samii Y (2005) Parallel particle swarm optimization and finite-difference time-domain (pso/fdtd) algorithm for multiband and wide-band patch antenna designs. IEEE T Antenn Propag 53:3459–3468. doi:10.1109/TAP.2005.858842
49. Johnson JM, Rahmat-Samii Y (1999) Genetic algorithms and method of moments (ga/mom) for the design of integrated antennas. IEEE T Antenn Propag 47:1606–1614. doi:10.1109/8.805906
50. Karimi-Moghaddam G, Gould R, Bhattacharya S (2012) Numerical investigation of electronic liquid cooling based on the thermomagnetic effect. In: Proceedings of the ASME 2012 international mechanical engineering Congress & Exposition, Houston, 9–15 Nov 2012. doi:10.1115/IMECE2012-87764

51. Kawamoto A, Matsumori T, Yamasaki S, Nomura T, Kondoh T, Nishiwaki S (2011) Heaviside projection based topology optimization by a pde-filtered scalar function. Struct Multidiscip O 44:19–24. doi:10.1007/s00158-010-0562-2
52. Kishk A, Ahn B, Kajfez D (1989) Broadband stacked dielectric-resonator antenna. Electron Lett 25:1232–1233. doi:10.1049/el:19890826
53. Kiziltas G, Psychoudakis D, Volakis JL, Kikuchi N (2003) Topology design optimization of dielectric substrates for bandwidth improvement of a patch antenna. IEEE T Antenn Propag 51:2732–2743. doi:10.1109/TAP.2003.817539
54. Kontoleontos EA, Papoutsis-Kiachagias EM, Zymaris AS, Papadimitriou DI, Giannakoglou KC (2012) Adjoint-based constrained topology optimization for viscous flows, including heat transfer. Eng Optimiz 45:941–961. doi:10.1080/0305215X.2012.717074
55. Lee J (2010) Structural design optimization of electric motors to improve torque performance. Dissertation, University of Michigan, Ann Arbor
56. Lee J, Kikuchi N (2010) Structural topology optimization of electrical machinery to maximize stiffness with body force distribution. IEEE T Magn 46:3790–3794. doi:10.1109/TMAG.2010.2052365
57. Lee J, Seo JH, Kikuchi N (2010) Topology optimization of switched reluctance motors for the desired torque profile. Struct Multidiscip O 42:783–796. doi:10.1007/s00158-010-0547-1
58. Lee J, Dede EM, Nomura T (2011) Simultaneous design optimization of permanent magnet, coils, and ferromagnetic material in actuators. IEEE T Magn 47:4712–4716. doi:10.1109/TMAG.2011.2160870
59. Lee J, Nomura T, Dede EM (2012) Heat flow control in thermo-magnetic convective systems using engineered magnetic fields. Appl Phys Lett 101:123507. doi:10.1063/1.4754119
60. Lee J, Nomura T, Dede EM (2012) Design optimization of magnetic fluid cooling system. In: Proceedings of the ASME 2012 international mechanical engineering Congress & Exposition, Houston, 9–15 Nov 2012. doi:10.1115/IMECE2012-85817
61. Leung KW, Luk KM, Yung EKN, Lai S (1995) Characteristics of a low-profile circular disk dr antenna with very high permittivity. Electron Lett 31:417–418. doi:10.1049/el:19950291
62. Lo H, Leung K, Luk K, Yung E (1999) Low-profile equilateral-triangular dielectric resonator antenna of very high permittivity. Electron Lett 35:2164–2166. doi:10.1049/el:19991459
63. Long SA, Conway GL, Shen L (1983) The resonant cylindrical dielectric cavity antenna. IEEE T Antenn Propag 31:406–412. doi:10.1109/TAP.1983.1143080
64. McAllister MW, Long SA, Conway GL (1983) Rectangular dielectric resonator antenna. Electron Lett 19:218–219. doi:10.1049/el:19830150
65. McAllister MW, Long SA (1984) Resonant hemispherical dielectric antenna. Electron Lett 20:657–659. doi:10.1049/el:19840450
66. Mlot A, Korkosz M, Lukaniszyn M (2012) Iron loss and eddy-current loss analysis in a low-power bldc motor with magnet segmentation. Arch Electr Eng 61:33–46. doi:10.2478/v10171-012-0003-5
67. Mongia RK, Ittipiboon A, Bhartia P, Cuhaci M (1993) Electric monopole antenna using a dielectric ring resonator. Electron Lett 29:1530–1531. doi:10.1049/el:19931019
68. Mongia RK, Ittipiboon A, Cuhaci M (1994) Low-profile dielectric resonator antennas using a very high permittivity material. Electron Lett 30:1362–1363. doi:10.1049/el:19940924
69. Nakatsuka K, Hama Y, Takahashi J (1990) Heat transfer in temperature-sensitive magnetic fluids. J Magn Magn Mater 85:207–209. doi:10.1016/0304-8853(90)90053-S
70. Narayana S, Sato Y (2012) Heat flux manipulation with engineered thermal materials. Phys Rev Lett 108:214303. doi:10.1103/PhysRevLett.108.214303
71. Nikolova NK, Tam HW, Bakr MH (2004) Sensitivity analysis with the fdtd method on structured grids. IEEE T Microw Theory 52:1207–1216. doi:10.1109/TMTT.2004.825710
72. Ning P, Ngo K, Wang F (2011) Thermomechanical reliability investigation of large temperature excursions in power electronics packages. In: Proceedings of the 2011 IEEE energy conversion Congress & Exposition, Phoenix, 17–22 Sept 2011. doi:10.1109/ECCE.2011.6063786
73. Nomura T, Sato K, Nishiwaki S, Yoshimura M (2007) Topology optimization of multiband dielectric resonator antennas. 7th World Congress on structural and multidisciplinary optimization, Seoul, 21–25 May 2007

74. Nomura T, Sato K, Taguchi K, Kashiwa T, Nishiwaki S (2007) Structural topology optimization for the design of broadband dielectric resonator antennas using the finite difference time domain technique. Int J Numer Meth Eng 71:1261–1296. doi:10.1002/nme.1974
75. Nomura T, Ohkado M, Schmalenberg P, Lee J, Ahmed O, Bakr M (2013) Topology optimization method for microstrips using boundary condition representation and adjoint analysis. European microwave conference (EuMC), Nuremberg, 6–10 Oct 2013
76. Nomura T, Yoon SW, Lee J, Dede EM (2013) Level set based topology optimization of directly bonded copper substrates targeting thermal stress minimization on die-substrate bonding line. 10th World Congress on structural and multidisciplinary optimization, Orlando, 19–24 May 2013
77. Olesen LH, Okkels F, Bruus H (2006) A high-level programming-language implementation of topology optimization applied to steady-state Navier-Stokes flow. Int J Numer Meth Eng 65:975–1001. doi:10.1002/nme.1468
78. Olszewski M (2011) Evaluation of the 2010 Toyota Prius hybrid synergy drive system. ORNL/TM-2010/253, Oak Ridge National Laboratory, Oak Ridge
79. Özişik MN (1993) Heat conduction, 2nd edn. Wiley, New York
80. Ozoe H (2005) Magnetic convection. Imperial College Press, London
81. Park S, Yoo J, Choi JS (2009) Simultaneous optimal design of the yoke and the coil in the perpendicular magnetic recording head. IEEE T Magn 45:3668–3671. doi:10.1109/TMAG.2009.2023878
82. Petosa A, Ittipiboon A, Antar Y, Roscoe D, Cuhaci M (1998) Recent advances in dielectric-resonator antenna technology. IEEE Antenn Propag M 40:35–48. doi:10.1109/74.706069
83. Reddy JN, Gartling DK (2000) The finite element method in heat transfer and fluid dynamics, 2nd edn. CRC Press, Boca Raton
84. Rosensweig RE (1985) Ferrohydrodynamics. Cambridge University Press, New York
85. Schwab L, Hildebrandt U, Stierstadt K (1983) Magnetic Bénard convection. J Magn Magn Mater 39:113–114. doi:10.1016/0304-8853(83)90412-2
86. Seo JH (2009) Optimal design of material microstructure for convective heat transfer in a solid-fluid mixture. Dissertation, University of Michigan, Ann Arbor
87. Sigmund O (2001) A 99 line topology optimization code written in Matlab. Struct Multidiscip O 21:120–127. doi:10.1007/s001580050176
88. Sobhan CB, Garimella SV (2001) A comparative analysis of studies on heat transfer and fluid flow in microchannels. Microscale Therm Eng 5:293–311. doi:10.1080/10893950152646759
89. Sullivan PF, Ramadhyani S, Incropera FP (1992) Extended surfaces to enhance impingement cooling with single circular round jets. Joint ASME/JSME conference on electronic packaging, Milpitas, 9–12 Apr 1992
90. Svanberg K (1987) The method of moving asymptotes—a new method for structural optimization. Int J Numer Meth Eng 24:359–373. doi:10.1002/nme.1620240207
91. Tsuji Y, Hirayama K, Nomura T, Sato K, Nishiwaki S (2006) Design of optical circuit devices based on topology optimization. IEEE Photonic Tech L 18:850–852. doi:10.1109/LPT.2006.871686
92. Tuckerman DB, Pease RFW (1981) High-performance heat sinking for VLSI. IEEE Electr Devices L 2:126–129. doi:10.1109/EDL.1981.25367
93. Yamasaki S, Nishiwaki S, Yamada T, Izui K, Yoshimura M (2010) A structural optimization method based on the level set method using a new geometry-based re-initialization scheme. Int J Numer Meth Eng 83:1580–1624. doi:10.1002/nme.2874
94. Yee K (1996) Numerical solution of initial boundary value problems involving Maxwell's equations in isotropic media. IEEE T Antenn Propag 14:302–307. doi:10.1109/TAP.1966.1138693
95. Yoon GH (2010) Topological design of heat dissipating structure with forced convective heat transfer. J Mech Sci Technol 24:1225–1233. doi:10.1007/s12206-010-0328-1
96. Zienkiewicz OC, Campbell JS (1973) Shape optimization and sequential linear programming. In: Gallagher RH, Zienkiewicz OC (eds) Optimum structural design. Wiley, New York, pp 109–126

Chapter 6
Extensions to New Topics

The field of multiphysics simulation is rapidly evolving due to continual increases in computing power and the multidisciplinary nature of engineering today. In this final chapter of the book, we collect our thoughts on how the numerical examples provided in this text may be extended to emerging ideas (or topics that have received limited attention) and that make use of state-of-the-art computational tools and optimization algorithms. While many of the case studies provided throughout this book incorporated two physical processes, only one example (related to thermomagnetic convection in Sect. 5.1.4) was focused on the simultaneous optimization of a system subject to three distinctly separate, yet coupled, physical phenomena. Here, we set forth a greater number of ideas related to these complex systems and leave it to the reader to explore them on their own. The relevance of such systems is that most actual electromechanical applications require the tight integration and handling of more than two physical processes in three dimensions, are highly constrained, and involve specific interface considerations.

6.1 Scaling-Up of Systems

A 2-D modeling approach was assumed for many of the multiphysics computational models and the majority of the optimization studies presented herein, in order to balance solution accuracy with overall computational cost. The validity of this assumption is critical to relevance of the numerical results, and it is clear that certain classes of problems would certainly benefit from the enriched design information that a 3-D modeling approach can provide. For example, in the case of conjugate heat transfer, performing an optimization study in two dimensions does little to resolve the 3-D effects of fin geometry on heat transfer enhancement.

In the case of structural optimization for multiphysics systems, the major obstacle in adopting a fully 3-D modeling strategy continues to be the excessive computational cost associated with obtaining design solutions. From a computational mechanics

E. M. Dede et al., *Multiphysics Simulation,* Simulation Foundations, Methods and Applications, DOI: 10.1007/978-1-4471-5640-6_6,

perspective, the co-design for many physical processes in a greater number of spatial dimensions equates to more degrees of freedom and larger systems of equations that must be solved. Thus, while it is expected that increased computing power (e.g., HPC systems) will assist in making this problem tractable, there exists an even greater need for better numerical techniques that enable the solution of large-scale multiphysics optimization problems in three dimensions. Some possible solutions to these large scale problems may include parallel processing for topology optimization [1, 6], the use of novel optimization schemes involving polyhedral meshes [8], or isogeometric analysis for topology optimization [5, 21, 24].

While moving from two to three dimensions in the spatial domain for multiphysics simulation and optimization opens new research areas, the handling of solutions beyond steady-state analysis must also be considered. The issue of dynamic response is of growing importance for higher efficiency electromechanical systems, see for example [13, 16], and the expansion of multiphysics design optimization tools into the temporal domain for transient analysis or frequency domain for eigenmode analysis is an area ripe for further exploration.

Additionally, most of the design examples presented in this book have dealt with linear material properties. However, one example in which this limiting assumption may be drawn into question is in relation to the evaluation of thermally induced stresses in electronics solder bond materials. It is well known that, when thermal cycled due to device power dissipation, solder bond attachment materials found in electronics can behave in a viscoplastic manner [29], where life cycle prediction is a function of cumulative damage. Therefore, the incorporation of more advanced nonlinear or temperature-dependent materials and appropriate modeling strategies [4] into multiphysics design optimization studies is critical to greater understanding of performance and reliability.

Building off of the discussion of linear versus non-linear materials, we have shown in Sect. 5.1.5 that the design of anisotropic materials and structures has the potential to open up new design spaces for certain classes of multiphysics problems. In the example in Sect. 5.1.5, a second-order tensor (i.e., the anisotropic material thermal conductivity tensor) was designed to manipulate the flow of heat in a composite structure. Logical extensions of this work to the design of structural systems involving higher order tensors is of relevance to both aerospace and automotive industries for composite design, and the co-design of structural topology and material layout, as proposed in [12, 17], represents another interesting area for future investigation.

Finally, another topic for greater development in the scaling-up of systems is the design optimization of multi-body electromechanical assemblies. Several of the actuator and motor examples described in Chap. 5 represent a starting point, and supplementing multi-body design approaches with the optimization of additional physical processes, richer 3-D models, and faster parallel solution techniques has great potential.

A graphical summary of the various above ideas related to the scaling-up of multiphysics simulation and design optimization tools and techniques is provided in the chart in Fig. 6.1.

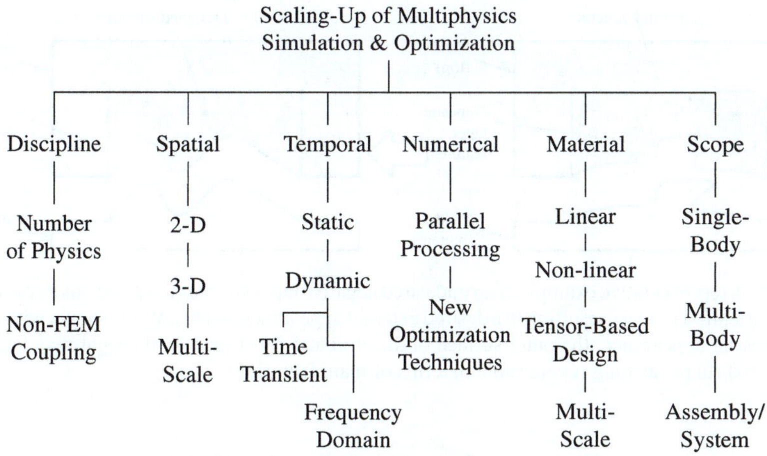

Fig. 6.1 Concepts related to the scaling-up of multiphysics simulation and design optimization tools and techniques

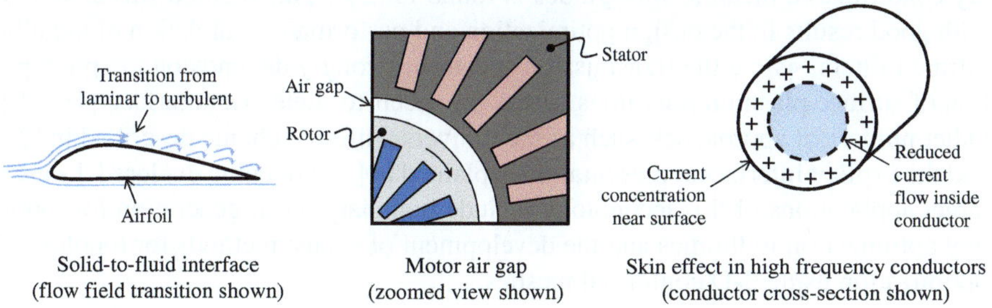

Fig. 6.2 Example problems for which the treatment of surfaces and interfaces is important in simulation and optimization

6.2 Treatment of Surfaces and Interfaces

One drawback of applying density-based topology optimization to the design of certain classes of multiphysics systems is the treatment of surfaces and interfaces, where the tracking of a boundary and assignment of appropriate model settings is important for arriving at accurate numerical solutions. Some examples include the resolution of boundary layers via a wall mesh in turbulent flow problems, modeling of two-phase (i.e., solid-to-liquid or liquid-to-vapor) heat transfer processes at material boundaries, mesh specification in the air gap region found in electrostatic actuators or between the rotor and stator in an electric motor assembly, and modeling of skin effects in high frequency conductors; refer to Fig. 6.2.

Given these challenges, robust simulation strategies are needed that allow for the proper tracking of boundaries during the optimization process. The level set function

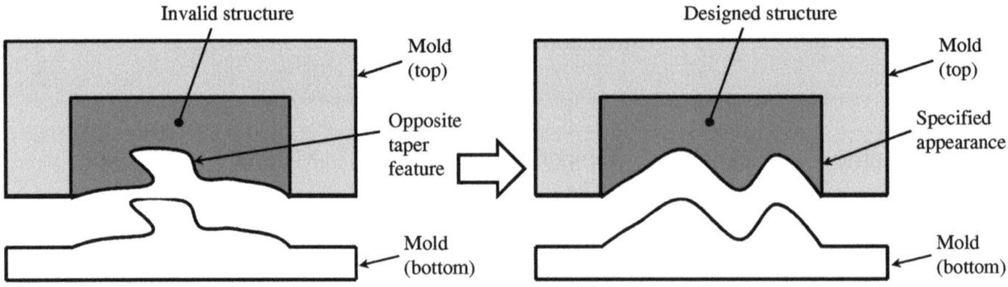

Fig. 6.3 A representative example of an undesired negative taper structure (*left*) versus a constrained mold fabrication process resulting in a desired external appearance (*right*). While shown with a '*box-like*' external appearance, the outer surface of the structure illustrated on the right may take most any desired shape, as long it is feasible in terms of manufacturing

approach outlined in Chap. 4, and recently reviewed in [27], has good potential for enabling advancements in this area. An early study related to boundary-tracking with an Arbitrary Lagrangian-Eulerian (ALE) mesh for imposing specific boundary conditions in metallic waveguides is found in [28]. This method was also used with good results in the design optimization and performance validation of metallic infrared filters, where the transmission spectrum strongly depends on the propagation of surface plasmon polaritons across the metal-to-dielectric interface [18, 19]. Other numerical approaches, such as the 'universal mesh' scheme described in [23] and the explicit interface representation explained in [3], are being explored. Logical target applications of this technology include boundary mesh generation for topology optimization in fluidics and the development of robust methods for topological optimization using 3-D conformal meshes.

6.3 Free Versus Constrained Systems-Toward Manufacturability

The scaling-up of systems paired with the treatment of important surfaces and interfaces is expected to provide a richer model definition to better capture and predict the behavior of electromechanical devices. Another feature that should be considered in conjunction with these enrichment strategies is the implementation of additional design constraints. Complex electromechanical systems are subject to performance constraints beyond volume or mass reduction [25], and as described in Sect. 5.3.2, fabrication constraints are often important to arrive at designs that are feasible from a manufacturing point of view; see Fig. 6.3.

An approach to this problem is to enforce a minimum length scale constraint using a projection method, as demonstrated in [9] for 2-D truss structural design. A further advanced example of this technique for smaller scale devices may be found in [20], where the authors mimic a MEMS fabrication process (see Fig. 6.4) with mathematical projection in the design of a process photomask for a structural rib; sample fabricated devices versus various optimal rib topologies are shown in Fig. 6.5

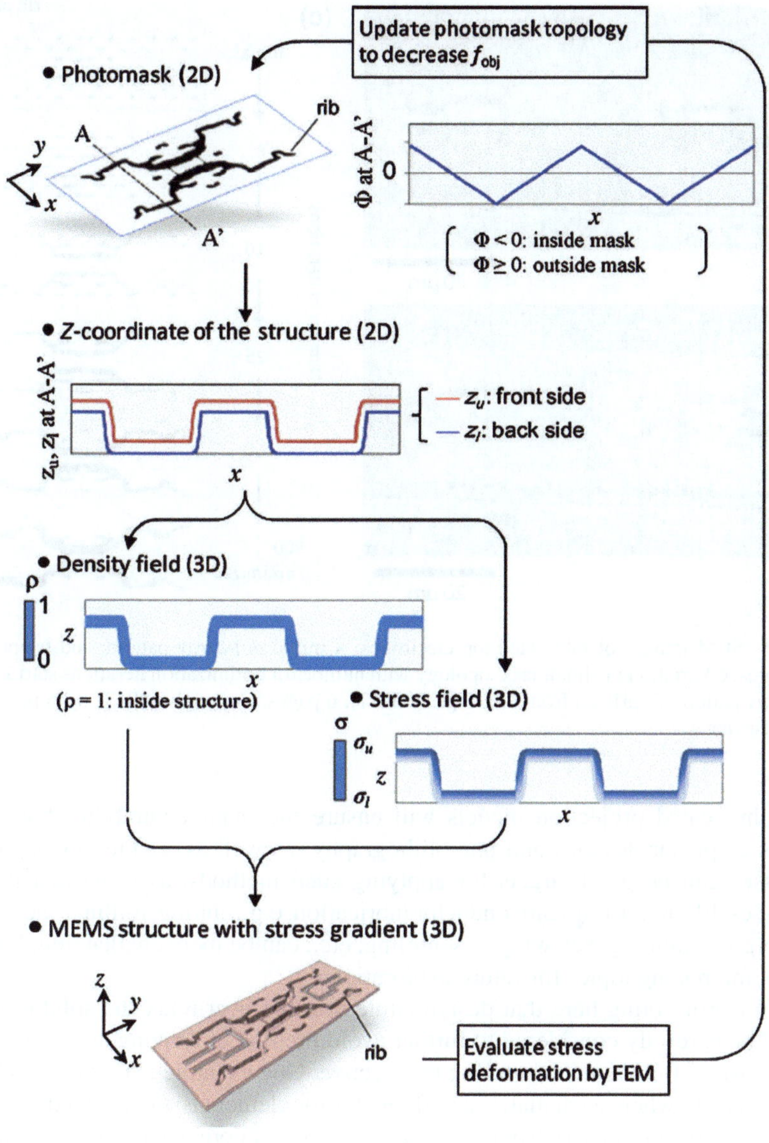

Fig. 6.4 Topology optimization flow using the micromachining process like mapping. Reprinted from [20, Fig. 3], Copyright (2013), with permission from Elsevier

on the left and right, respectively. Note that a similar strategy was used in Sect. 5.1.2, where a 2-D etching process photomask pattern was optimized and projected into 3-D for the design of a larger scale DBC substrate that minimized the thermally induced stresses present at the device bonding layer in an electronics package.

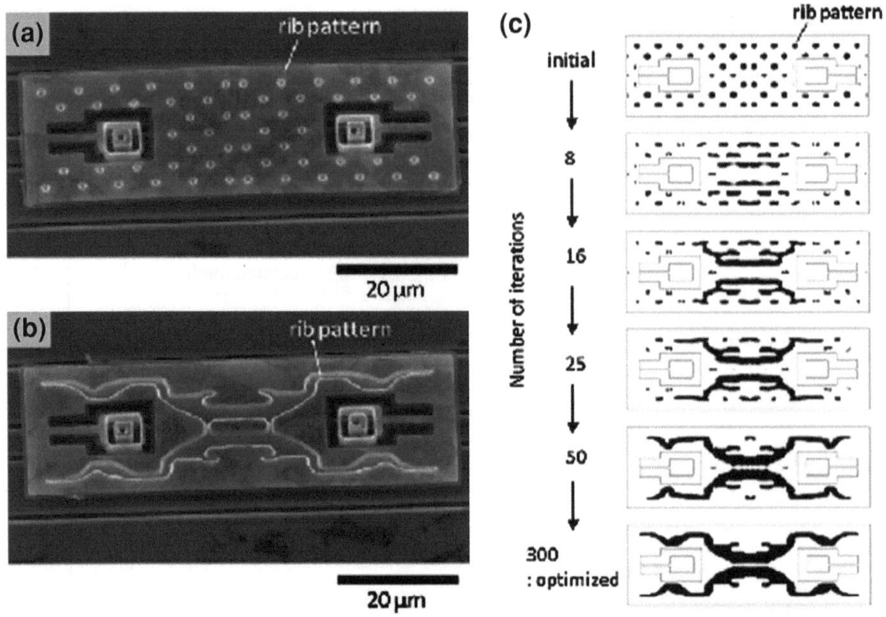

Fig. 6.5 SEM images of fabricated top electrodes: **a** initial *dot-array* pattern, and **b** optimized topologies. **c** Variation in photomask topology with number of optimization iterations starting from *dot-array* pattern. Reprinted from [20, Figs. 6 and 7], 6 pages, Copyright (2013), with permission from Elsevier

Sophisticated projection models will ensure the manufacturability for certain classes of planar designs, and photolithography, as well as, etching or deposition processes can be good targets for applying such methods, as suggested in [11]. Processes different from semiconductor fabrication, e.g., cutting, milling, sheet forming, forging, welding, screwing, assembling, etc., can be more challenging, but they remain interesting topics for future exploration.

It is worth noting here that design strategies that over-relax the solution space should be carefully considered to further avoid the risk of arriving at solutions that are infeasible from a manufacturing perspective. One example is free material optimization [10], where each material stiffness tensor element is an independent design variable. While the approach may be attractive from a theoretical perspective, the feasibility of the existence of the material microstructure is lost and difficult to recover, and it may be more practical to constrain the design of the material or structure to solutions that are possible in terms of physical implementation.

Finally, as the field of optimization of multiphysics electromechanical devices in three dimensions expands, the need for efficient CAD techniques for model geometry synthesis will also increase. The efficient extraction, smoothing, and handling of point cloud data in an automated fashion for the generation of solid model CAD geometry, while preserving object topology for subsequent design, analysis,

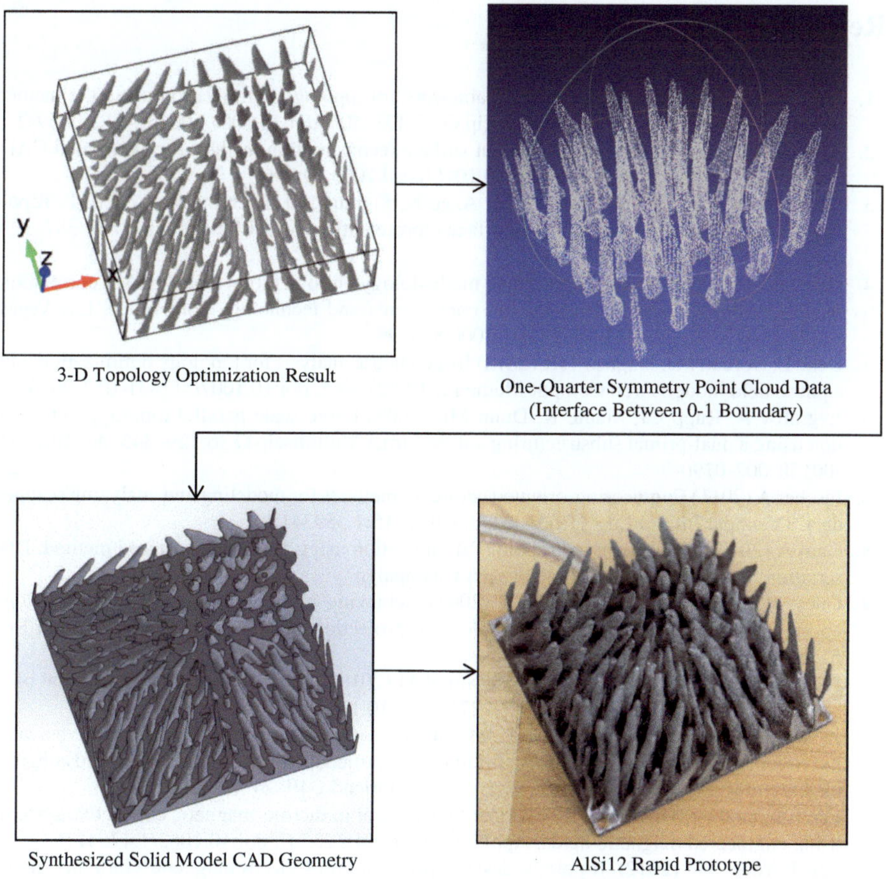

Fig. 6.6 Representative process for synthesizing model geometry—from optimization to rapid prototype. The figure shows a pin fin heat sink optimized in 3-D for air jet impingement cooling

or manufacturing [2, 7] is an interesting area that deserves greater attention due to logical intersections with the field of 3-D structural design optimization. Figure 6.6 illustrates a representative design process flow for the model geometry synthesis of an air cooled pin fin heat sink optimized in three dimensions for jet impingement cooling. Inherently, this type of 3-D topology optimization requires automated technology to advance from the research and development stage to the design engineering stage. The coupling of these geometry synthesis tools with robust design optimization strategies [22, 26] and next generation computer numerically controlled (CNC) multi-axis machining or additive manufacturing methods [14, 15] offers significant potential as a set of highly disruptive technologies that enable ultra-efficient electromechanical device designs.

References

1. Aage N, Lazarov BS (2013) Parallel framework for topology optimization using the method of moving asymptotes. Struct Multidiscip O 47:493–505. doi:10.1007/s00158-012-0869-2
2. Azernikov S, Fischer A (2004) Efficient surface reconstruction method for distributed CAD. Comput Aided Des 36:799–808. doi:10.1016/j.cad.2003.09.006
3. Christiansen AN, Nobel-Jørgensen M, Aage N, Sigmund O, Baerentzen JA, (2014) Topology optimization using an explicit interface representation. Struct Multidiscip O 49:387–399. doi:10.1007/s00158-013-0983-9
4. Darveaux R (2000) Effect of simulation methodology on solder joint crack growth correlation. In: Proceedings of the 50th electronic components and technology conference, Las Vegas, 21–24 May 2000. doi:10.1109/ECTC.2000.853299
5. Dedè L, Borden MJ, Hughes TJR (2012) Isogeometric analysis for topology optimization with a phase field model. Arch Comput Method E 19:427–465. doi:10.1007/s11831-012-9075-z
6. Evgrafov A, Rupp CJ, Maute K, Dunn ML (2008) Large-scale parallel topology optimization using a dual-primal substructuring solver. Struct Multidiscip O 36:329–345. doi:10.1007/s00158-007-0190-7
7. Fischer A (2011) Engineering-oriented geometry methods for modeling and analyzing scanned data. J Comput Inf Sci Eng 11:021002. doi:10.1115/1.3593415
8. Gain A (2013) Polytope-based topology optimization using a mimetic-inspired method. Dissertation, University of Illinois at Urbana-Champaign
9. Guest JK, Prévost JH, Belytschko T (2004) Achieving minimum length scale in topology optimization using nodal design variables and projection functions. Int J Numer Meth Eng 61:238–254. doi:10.1002/nme.1064
10. Haslinger J, Kočvara M, Leugering G, Stingl M (2010) Multidisciplinary free material optimization. SIAM J Appl Math 70:2709–2728. doi:10.1137/090774446
11. Jansen M, Lazarov BS, Schevenels M, Sigmund O (2013) On the similarities between micro/nano lithography and topology optimization projection methods. In: 10th world congress on structural and multidisciplinary optimization, Orlando, 19–24 May 2013
12. Lee J, Nomura T, Dede EM (2012) Heat flow control in thermo-magnetic convective systems using engineered magnetic fields. Appl Phys Lett 101:123507. doi:10.1063/1.4754119
13. Lee J, Yoon SW (2012) Topology design optimization of electromagnetic vibration energy harvester to maximize power output. J Magn 18:283–288. doi:10.4283/JMAG.2013.18.3.283
14. Lipson H, Kurman M (2013) Fabricated: the new world of 3D printing. Wiley, Indianapolis
15. Malone E, Lipson H (2008) Multi-material freeform fabrication of active systems. In: Proceedings of the ASME 2008 9th biennial conference on engineering systems design and analysis, Haifa, 7–9 July 2008. doi:10.1115/ESDA2008-59313
16. Noh JY, Yoon GH (2012) Topology optimization of piezoelectric energy harvesting devices considering static and harmonic dynamic loads. Adv Eng Softw 53:45–60. doi:10.1016/j.advengsoft.2012.07.008
17. Nomura T, Dede EM, Lee J, Yamasaki S, Matsumori T, Kawamoto A, Kikuchi N (2014) General topology optimization method with continuous and discrete orientation design using isoparametric projection (submitted)
18. Ohkado M, Nomura T, Matsumori T, Kawamoto A, Fujikawa H, Sato K, Yamasaki S, Nishiwaki S (2011) Structural optimization of SPPs color filter using a level set-based topology optimization method incorporating the ALE method. In: 9th world congress on structural and multidisciplinary optimization, Shizuoka, 13–17 June 2011
19. Ohkado M, Nomura T, Miura A, Fujikawa H, Ikeda N, Sugimoto Y, Nishiwaki S (2013) Structural optimization of metallic infrared filters based on extraordinary optical transmission. T MRS Jap 38:167–170. doi:10.14723/tmrsj.38.167
20. Ozaki T, Nomura T, Fujitsuka N, Shimaoka K, Akashi T (2013) Topology optimization using multistep mapping from 2D photomask to 3D structure for designing reinforcing rib. Sensors Actuat A-Phys (in press). doi:10.1016/j.sna.2013.08.033

21. Qian X (2013) Topology optimization in B-spline space. Comput Method Appl M 265:15–35. doi:10.1016/j.cma.2013.06.001
22. Qian X, Sigmund O (2013) Topological design of electromechanical actuators with robustness toward over- and under-etching. Comput Method Appl M 253:237–251. doi:10.1016/j.cma.2012.08.020
23. Rangarajan R, Lew AJ (2012) Universal meshes: a new paradigm for computing with nonconforming triangulations. http://arxiv.org/abs/1201.4903v1. Accessed 19 March 2014
24. Seo YD, Kim HJ, Youn SK (2010) Isogeometric topology optimization using trimmed spline surfaces. Comput Method Appl M 199:3270–3296. doi:10.1016/j.cma.2010.06.033
25. Sigmund O (2001) Design of multiphysics actuators using topology optimization-Part I: one-material structures. Comput Method Appl M 190:6577–6604. doi:10.1016/S0045-7825(01)00251-1
26. Sigmund O (2009) Manufacturing tolerant topology optimization. Acta Mech Sinica 25:227–239. doi:10.1007/s10409-009-0240-z
27. van Dijk NP, Maute K, Langelaar M, van Keulen F (2013) Level-set methods for structural topology optimization: a review. Struct Multidiscip O 48:437–472. doi:10.1007/s00158-013-0912-y
28. Yamasaki S, Nomura T, Kawamoto A, Sato K, Nishiwaki S (2011) A level set-based topology optimization method targeting metallic waveguide design problems. Int J Numer Meth Eng 87:844–868. doi:10.1002/nme.3135
29. Ye H, Lin M, Basaran C (2002) Failure modes and FEM analysis of power electronic packaging. Finite Elem Anal Des 38:601–612. doi:10.1016/S0168-874X(01)00094-4

Chapter 7
Appendix: Sample Multiphysics Optimization Code

In this appendix, example numerical code is provided for a 2-D electrothermal design optimization problem similar to the example presented in Sect. 5.1.1. Here, Joule heating considering side/surface convection (as described in [3]) of the conductor is once more assumed. However, in this case, the solid/void surface convection boundary condition is interpolated using the first derivative of the logistic function, for simplicity, instead of the smoothed hat-function, Eq. (5.7).

The finite element implementation of the problem follows the description in [1] for the finite element analysis (FEA) of a thermoelectric device, with the exception that Seebeck and Peltier effects are neglected. The code is written in a custom MATLAB® [1] script that exploits a sequential solver for the FEA portion of the problem, where the electrical analysis is first performed and the generated resistive heating thermal loads are passed to the following heat transfer analysis. The code incorporates the OC optimizer and filtering routine provided in [4] for minimum length scale control. The code is constructed as a continuous script instead of a series of functions for clarity, and numerous comments have been added throughout the script to explain the various steps implemented in the program.

A 2-D schematic of the problem set up is provided in Fig. 7.1, where a square design domain is specified with a current source assumed at Terminal 1 (top left) plus electrical ground and a fixed temperature condition at Terminal 2 (bottom right). Material properties of copper were assumed for both the material electrical conductivity and thermal conductivity. Further interpretation of the sample code is left to the reader, and additional FEA implementation details may be found in [2]. The reader is also referred to Sect. 5.1.1 for a slightly modified, yet related, example problem that was implemented instead in a commercial software environment.

[1] MATLAB® is a registered trademark of The MathWorks, Inc.

E. M. Dede et al., *Multiphysics Simulation,* Simulation Foundations, Methods and Applications, DOI: 10.1007/978-1-4471-5640-6_7,

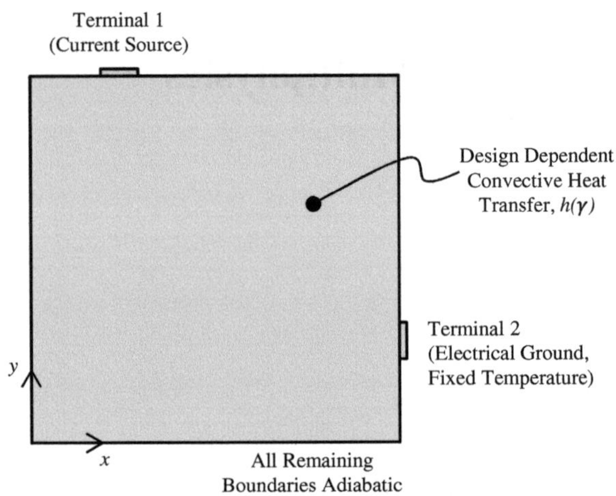

Fig. 7.1 Assumed 2-D simulation domain for electrothermal optimization example written in MATLAB® programming language. A current source is assumed at Terminal 1 with electrical ground and thermal ground at Terminal 2. Design dependent convective heat transfer, $h(\gamma)$, is assumed inside the domain with all remaining external boundaries considered adiabatic

7.1 MATLAB® Example Program for Multiphysics Topology Optimization of Electrothermal Systems

```
%-----------------------------------------------------------------
% Multiphysics Topology Optimization Code
% Written by: E.M. Dede, T. Nomura, J. Lee
% Last Modified: April 1st, 2014
% Description: This code optimizes a 2-D electrothermal system
% with current input to the top left terminal of the square
% domain and electrical/thermal ground at the bottom right
% terminal. The electrical power is used as the internal heat
% generation for the thermal portion. The electrical
% conductivity, thermal conductivity, and surface convection
% coefficient are interpolated on the domain. A sequential FEA
% solution process is assumed. An OC optimizer is used
%-----------------------------------------------------------------
clc                                % Clear the workspace
clear all                          % Clear the variables

maxiter = 100;                     % Max # of iterations

nelx = 50.0;                       % # of elements in x-dir
nely = 50.0;                       % # of elements in y-dir
volfrac = 0.2;                     % Define volume fraction
```

```
penal = 3.0;                        % Penalization power
rmin = 1.05;                        % Filter radius

%--------------------------------------------------------------
% Try w1 values of 1, 0.98, 0.95, 0.9, 0.78, 0.5, and 0.25
w1 = 1;                             % Electrical weight
w2 = 1-w1;                          % Thermal weight
%--------------------------------------------------------------

% Variables for 'h' boundary condition interpolation
lfp = 15;          % Logistic function bandwidth parameter
lfoff = 0.6;       % Logistic function offset
ho = 2.5;          % Assumed convection coefficient value
Tamb = 293;        % Ambient reference temperature

xsiz = 0:1:nelx;% Define meshgrid for plotting
ysiz = 0:1:nely;
[xx,yy] = meshgrid(xsiz,ysiz);

% Initialize design variables
x(1:nely,1:nelx) = volfrac;
                   % Distribute material evenly in design domain
loop = 0;          % Initialize main loop
change = 1.;       % Initialize change in design variables

% Start optimization loop
tic
%--------------------------------------------------------------

while loop < maxiter
    loop = loop + 1;       % Increment loop counter
    xold = x;              % Set old design variable matrix

    % Electrical FEA
    %----------------------------------------------------------
    sigma = 3.5E7;         % Electrical conductivity
    KEe = sigma*[ 2/3 -1/6 -1/3 -1/6
        -1/6 2/3 -1/6 -1/3
        -1/3 -1/6 2/3 -1/6
        -1/6 -1/3 -1/6 2/3];

    % Initialize electrical system K matrix plus global current
    % load and voltage displacement vectors
    Ke = sparse((nelx+1)*(nely+1), (nelx+1)*(nely+1));
    Fe = sparse((nely+1)*(nelx+1),1);
```

```
Ue = sparse((nely+1)*(nelx+1),1);

% Loop to form the global stiffness matrix
for ely = 1:nely              % Loop over # el y-dir.
    for elx = 1:nelx          % Loop over # el x-dir.
        % Upper left el node number in GCS
        n1 = (nely+1)*(elx-1)+ely;
        % Upper right el node number in GCS
        n2 = (nely+1)* elx   +ely;
        % Element dof
        edof = [n1; n2; n2+1; n1+1];
        % Populate electrical system K matrix
        Ke(edof,edof) = Ke(edof,edof) + ...
            (0.001+0.999*x(ely,elx)^penal)*KEe;
    end
end


% Define electrical loads and boundary conditions
% Apply current at top left terminal
Fe([10*(nely+1)+1,11*(nely+1)+1,...
    12*(nely+1)+1,13*(nely+1)+1,...
    14*(nely+1)+1],1) = 2000;
% Lower right terminal is electrical ground
fixeddofs = ...
    [((nelx+1)*(nely+1)-15):((nelx+1)*(nely+1)-11)];
alldofs   = [1:(nely+1)*(nelx+1)];
freedofs  = setdiff(alldofs,fixeddofs);

% Solve for the voltage state variables
Ue(freedofs,:) = Ke(freedofs,freedofs) \ Fe(freedofs,:);
Ue(fixeddofs,:)= 0;

% Distribute voltage state variables to array
p = zeros(nely+1,nelx+1);
 for i = 1:nelx+1
     for j = 1:nely+1
         p(j,i) = Ue((i-1)*(nely+1)+j);
     end
 end

 % Solve for gradients of potential field
 [px,py] = gradient(p,1,1);

 % Compute the electrical power
```

```
Qe=sigma*(px.^2+py.^2);

% Thermal FEA
%-----------------------------------------------------------
k = 400;                          % Thermal conductivity
KEt = k*[ 2/3 -1/6 -1/3 -1/6% Conductivity matrix
    -1/6 2/3 -1/6 -1/3
    -1/3 -1/6 2/3 -1/6
    -1/6 -1/3 -1/6 2/3];

KEh = ho*[ 2/3 1/6  0   1/6 % Convection matrix
    1/6 2/3  1/6 0
    0   1/6  2/3 1/6
    1/6 0    1/6 2/3];
FEh = ho*Tamb;                    % Convective loads

% Initialize thermal system K matrix plus global thermal
% load, convective load, and temperature displacement
% vectors
Kt = sparse((nelx+1)*(nely+1), (nelx+1)*(nely+1));
Ft = sparse((nely+1)*(nelx+1),1);
Fh = sparse((nely+1)*(nelx+1),1);
Ut = sparse((nely+1)*(nelx+1),1);

% Loop to form the global stiffness matrix
for ely = 1:nely                  % Loop over # el y-dir.
    for elx = 1:nelx              % Loop over # el x-dir.
        % Upper left el node number in GCS
        n1 = (nely+1)*(elx-1)+ely;
        % Upper right el node number in GCS
        n2 = (nely+1)* elx   +ely;
        % Element dof
        edof = [n1; n2; n2+1; n1+1];
        % Logistic function for 'h' interpolation
        lfi = (1/(1+exp(-lfp*(x(ely,elx)-lfoff))));
        % Populate thermal K matrix
        Kt(edof,edof) = Kt(edof,edof) + ...
            (0.001+0.999*x(ely,elx)^penal)*KEt + ...
            (lfi*(1-lfi))*KEh;
    end
end


% Define thermal loads and boundary conditions
% Add thermal loads from electrical problem
```

```
for i = 1:nelx+1
    for j = 1:nely+1
        Ft((i-1)*(nely+1)+j,1) = Qe(j,i);
    end
end


% Compute and add in convective loads
for ely = 1:nely
    for elx = 1:nelx
        % Upper left el node number in GCS
        n1 = (nely+1)*(elx-1)+ely;
        % Upper right el node number in GCS
        n2 = (nely+1)* elx   +ely;
        % Element dof
        edof = [n1; n2; n2+1; n1+1];
        % Logistic function for 'h' interpolation
        lfi = (1/(1+exp(-lfp*(x(ely,elx)-lfoff))));
        % Determine convective loads
        Fh(edof,1) = Fh(edof,1) + (lfi*(1-lfi))*FEh;
    end
end

%Add convective loads
Ft = Ft + Fh;

% Lower right terminal is set to zero temperature
fixeddofs = [((nelx+1)*(nely+1)-15):((nelx+1)*(nely+1)-11)];
alldofs   = [1:(nely+1)*(nelx+1)];
freedofs  = setdiff(alldofs,fixeddofs);

% Solve for the temperature state variables
Ut(freedofs,:) = Kt(freedofs,freedofs) \ Ft(freedofs,:);
Ut(fixeddofs,:)= 0;

% Evaluate the objective function and perform sensitivity
% analysis
%------------------------------------------------------------
c = 0.;          % Initialize the objective function
for ely = 1:nely                    % Loop # el y-dir.
    for elx = 1:nelx                % Loop # el x-dir.
        % Upper left el node number in GCS
        n1 = (nely+1)*(elx-1)+ely;
        % Upper right el node number in GCS
        n2 = (nely+1)* elx   +ely;
```

```
            % Extract el node displacements
            UEe = Ue([n1; n2; n2+1; n1+1],1);
            UEt = Ut([n1; n2; n2+1; n1+1],1);
            % Compute objective function
            c = c + (0.001+0.999*x(ely,elx)^penal)*...
                (w1*UEe'*KEe*UEe + w2*UEt'*KEt*UEt);
            % Compute sensitivities
            dc(ely,elx) = ...
                -0.999*penal*x(ely,elx)^(penal-1)*...
                (w1*UEe'*KEe*UEe + w2*UEt'*KEt*UEt);
        end
    end


    % Filtering of sensitivities
    %----------------------------------------------------------------
    % Initialize modified sensitivity matrix
    dcn=zeros(nely,nelx);
    for i=1:nelx                        % Loop # el x-dir.
        for j=1:nely                    % Loop # el y-dir.
            sum=0.0;
            for k = max(i-round(rmin),1):...
                    min(i+round(rmin),nelx)
                for l = max(j-round(rmin),1):...
                        min(j+round(rmin),nely)
                    % Compute conv. factor
                    fac = rmin-sqrt((i-k)^2+(j-l)^2);
                    sum = sum+max(0,fac);
                    % Modified sensitivity
                    dcn(j,i) = dcn(j,i) + ...
                        max(0,fac)*x(l,k)*dc(l,k);
                end
            end
             % Finish compute modifying sensitivities
            dcn(j,i)=dcn(j,i)/(x(j,i)*sum);
        end
    end
    dc=dcn;


    % Design update by the optimality criteria method
    %----------------------------------------------------------------
    % Bounds for lagrange multi. & move-limit
    l1=0; l2=100000; move=0.2;
    % Lagrange multiplier convergence criteria
```

```
    while (l2-l1 > 1e-4)
        lmid = 0.5*(l2+l1); % Find mid-point of l1 & l2
        xnew = ...           % Optimality criteria
            max(0.001,max(x-move,min(1.,min(x+...
            move,x.*sqrt(-dc./lmid)))));
        xsum = 0;
        % Sum over xnew
        for i = 1:nelx
            for j = 1:nely
                xsum = xsum + xnew(j,i);
            end
        end
        % Bi-section algorithm for Lagrange multiplier
        if xsum - volfrac*nelx*nely > 0;
            l1=lmid;
        else
            l2=lmid;
        end
    end
    x = xnew;


    % Print topology results
    %------------------------------------------------------------
    change = max(max(abs(x-xold)));
    disp([' It.: ' sprintf('%4i',loop) ...
        ' Obj.: ' sprintf('%10.4f',c) ...
        ' Vol.: ' sprintf('%6.3f',xsum/(nelx*nely))...
        ' ch.: ' sprintf('%6.3f',change) ])

    %Plot the density
    colormap(gray); imagesc(-x); axis equal; ...
        axis tight; axis off; pause(1e-8);
    title('Density [Dark Regions = Solid; Light Regions = Void]')

end                                  % Loop for iterations
toc


% Post-processing
%----------------------------------------------------------------
% Plot the electric potential
Ueplot = zeros(nely+1,nelx+1);
for i = 1:nelx+1
    for j = 1:nely+1
```

```
        Ueplot(j,i) = Ue((i-1)*(nely+1)+j);
    end
end
figure; imagesc(Ueplot); axis equal; axis tight; axis off; colorbar
title('Electric Potential [SI Units: V]')


% Plot the E-field
figure; quiver(xx,nely-yy,-px,py,3); ...
    axis equal; axis tight; axis off
title('Electric Field Vector Plot')

% Plot the electrical power
figure; imagesc(Qe); axis equal; axis tight; axis off; colorbar
title('Power [SI Units: W]')

% Plot the temperature field
Utplot = zeros(nely+1,nelx+1);
for i = 1:nelx+1
    for j = 1:nely+1
        Utplot(j,i) = Ut((i-1)*(nely+1)+j);
    end
end
figure; imagesc(Utplot); axis equal; axis tight; axis off; colorbar
title('Temperature [SI Units: K]')

% Plot the 'h' field
Fhplot = zeros(nely+1,nelx+1);
for i = 1:nelx+1
    for j = 1:nely+1
        Fhplot(j,i) = Fh((i-1)*(nely+1)+j)/Tamb;
    end
end
figure; imagesc(Fhplot); axis equal; axis tight; axis off; colorbar
title('Surface Convection Coefficient [SI Units: W/(m^2*K)]')
```

References

1. Antonova EE, Looman DC (2005) Finite elements for thermoelectric device analysis in ANSYS. Paper presented at the 24th international conference on thermoelectrics, Clemson University, Clemson, 19–23. doi:10.1109/ICT.2005.1519922
2. Kwon YW, Bang H (2000) The finite element method using MATLAB, 2nd edn. CRC Press, Boca Raton
3. Seo JH (2009) Optimal design of material microstructure for convective heat transfer in a solid-fluid mixture. Dissertation, University of Michigan, Ann Arbor
4. Sigmund O (2001) A 99 line topology optimization code written in Matlab. Struct Multidiscip O 21:120–127. doi:10.1007/s001580050176

Index

E. M. Dede et al., *Multiphysics Simulation*, Simulation Foundations, Methods and Applications, DOI: 10.1007/978-1-4471-5640-6,

MIX
Papier aus verantwortungsvollen Quellen
Paper from responsible sources
FSC® C105338

If you have any concerns about our products,
you can contact us on
ProductSafety@springernature.com

In case Publisher is established outside the EU,
the EU authorized representative is:
Springer Nature Customer Service Center GmbH
Europaplatz 3, 69115 Heidelberg, Germany

Printed by Libri Plureos GmbH
in Hamburg, Germany